METHODS IN NONLINEAR INTEGRAL EQUATIONS

Methods in Nonlinear Integral Equations

by

Radu Precup
Department of Applied Mathematics,
Babeş-Bolyai University,
Cluj, Romania

KLUWER ACADEMIC PUBLISHERS
DORDRECHT / BOSTON / LONDON

A C.I.P. Catalogue record for this book is available from the Library of Congress.

ISBN 1-4020-0844-9

Published by Kluwer Academic Publishers,
P.O. Box 17, 3300 AA Dordrecht, The Netherlands.

Sold and distributed in North, Central and South America
by Kluwer Academic Publishers,
101 Philip Drive, Norwell, MA 02061, U.S.A.

In all other countries, sold and distributed
by Kluwer Academic Publishers,
P.O. Box 322, 3300 AH Dordrecht, The Netherlands.

Printed on acid-free paper

Printed in the Netherlands.

To Rodica and Barbu with love

Contents

Preface

Nonlinear integral equations represent the major source of nonlinear operators and, by this, equally, the motor and the testing ground of a vast domain of nonlinear analysis, namely the theory of nonlinear operators. This book is intended to present some of the most useful tools of nonlinear analysis in studying systems of integral equations (integral equations in $\mathbf{R}^n$).

The book is divided into three parts: fixed point methods, variational methods, and iterative methods. Each part contains a number of chapters of theory and applications. The methods we deal with are: Schauder's fixed point theorem; the Leray–Schauder principle; direct variational methods; the mountain pass theorem; the discrete continuation principle; monotone iterative techniques; methods of upper and lower solutions; Newton's method; and the generalized quasilinearization method.

We note that the selected topics reflect the particular interests of the author and that many contributions to the subject, as well as some other important treatment methods of the nonlinear integral equations have not been included for the sake of brevity.

The lists of references, one for each part, include only referenced titles; they are just intended to guide the reader through the enormous existing literature and not to provide a complete bibliography on the subject.

The presentation is essentially self-contained and leads the reader from basic concepts and results to current ideas and methods of nonlinear analysis.

We hope the book will be of interest to graduate students, and theoretical and applied mathematicians in nonlinear functional analysis, integral equations, ordinary and partial differential equations and related fields.

Radu Precup

Notation

$\mathbf{R}_+$	set of all nonnegative real numbers
$\mathbf{R}^n$	set of all n-tuples $x = (x_1, x_2, ..., x_n)$
$\mathbf{R}^n_+$	set of all $x \in \mathbf{R}^n$ with $x_i \geq 0$ for all i
$\lvert x\rvert$	Euclidian norm of $x \in \mathbf{R}^n$, $\lvert x\rvert = \left(\sum_{i=1}^n x_i^2\right)^{1/2}$
(x, y)	Euclidian inner product in $\mathbf{R}^n$, $(x, y) = \sum_{i=1}^n x_i y_i$, also denoted by $x \cdot y$
$x \leq y$	natural order relation in $\mathbf{R}^n$: $x_i \leq y_i$ for all i
$x < y$	strict order relation in $\mathbf{R}^n$: $x_i < y_i$ for all i
$x \leq a$	for $x \in \mathbf{R}^n$ and $a \in \mathbf{R}$: $x_i \leq a$ for all i
μ	Lebesgue measure in $\mathbf{R}^n$
X, Y, Z	metric spaces, real Banach or Hilbert spaces
X^*	dual space of X
$\lvert . \rvert$	norm in X, also denoted by $\lvert . \rvert_X$
$(.,.)$	inner product in a Hilbert space; also for $u \in X$ and $u^* \in X^*$, (u^*, u) is the value of u^* at u, $u^*(u)$
$B_r(u; X)$	open ball $\{v \in X : \lvert u - v\rvert < r\}$ ($B_r(u)$ for short)
$\overline{B}_r(u; X)$	closed ball $\{v \in X; \lvert u - v\rvert \leq r\}$ ($\overline{B}_r(u)$ for short)
$\overline{U}$, $\operatorname{int} U$	closure of U, interior of U
∂U	boundary of U : $\partial U = \overline{U} \setminus \operatorname{int} U$
$\operatorname{conv} A$	convex hull of A
$\overline{\operatorname{conv}} A$	closed convex hull of A
$C^k(\Omega; \mathbf{R}^n)$	set of k-times continuously differentiable functions $u : \Omega \to \mathbf{R}^n$ ($\Omega \subset \mathbf{R}^N$ open)
$C^k(\overline{\Omega}; \mathbf{R}^n)$	space of all functions $u \in C^k(\Omega; \mathbf{R}^n)$, $u = (u_1, ..., u_n)$ such that $D^\alpha u_i$ admits a continuous extension to $\overline{\Omega}$ for all i and $\alpha = (\alpha_1, ..., \alpha_N)$ with $\lvert\alpha\rvert = \sum_{j=1}^N \alpha_j \leq k$. Here $D^\alpha = \partial^{\lvert\alpha\rvert} / \partial x_1^{\alpha_1} ... \partial x_N^{\alpha_N}$

$\lvert u\rvert_\infty$	$\max_{x\in\overline{\Omega}} \lvert u(x)\rvert$ ($\Omega \subset \mathbf{R}^N$ bounded open, $u \in C(\overline{\Omega};\mathbf{R}^n)$)
$C^k(\overline{\Omega})$	stands for $C^k(\overline{\Omega};\mathbf{R})$
$C^k[a,b]$	stands for $C^k([a,b])$
$L^p(\Omega;\mathbf{R}^n)$	space of all measurable functions $u:\Omega\to\mathbf{R}^n$ with $\int_\Omega \lvert u(x)\rvert^p\,dx < \infty$ ($\Omega\subset\mathbf{R}^N$ open, $1\le p<\infty$)
$\lvert . \rvert_p$	norm in $L^p(\Omega;\mathbf{R}^n)$, $\lvert u\rvert_p = \left(\int_\Omega \lvert u(x)\rvert^p\,dx\right)^{1/p}$
$(.,.)_2$	inner product in $L^2(\Omega;\mathbf{R}^n)$, $(u,v)_2 = \int_\Omega (u(x),v(x))\,dx$
$L^\infty(\Omega;\mathbf{R}^n)$	space of all measurable functions $u:\Omega\to\mathbf{R}^n$ for which there is a constant c with $\lvert u(x)\rvert\le c$ for a.e. $x\in\Omega$
$\lvert . \rvert_\infty$	norm in $L^\infty(\Omega;\mathbf{R}^n)$, $\lvert u\rvert_\infty = \inf\{c: \lvert u(x)\rvert \le c \text{ a.e. on } \Omega\}$
$L^p(\Omega)$	stands for $L^p(\Omega;\mathbf{R})$ $(1\le p\le\infty)$
$L^p(a,b;\mathbf{R}^n)$	stands for $L^p(\Omega;\mathbf{R}^n)$ with $\Omega=(a,b)$
$L^p(a,b)$	stands for $L^p(a,b;\mathbf{R})$
$L^p(\Omega;X)$	space of all stongly measurable functions $u:\Omega\to X$ with $\lvert u\rvert_X \in L^p(\Omega)$

Basic inequality:

$$\left|\int_\Omega u(x)\,dx\right| \le \int_\Omega \lvert u(x)\rvert\,dx, \quad u\in L^1(\Omega;\mathbf{R}^n),$$

where $\Omega\subset\mathbf{R}^N$ is open, $\lvert . \rvert$ is the Euclidian norm in $\mathbf{R}^n$, and

$$\int_\Omega u(x)\,dx = \left(\int_\Omega u_1(x)\,dx, \ldots, \int_\Omega u_n(x)\,dx\right).$$

Here $u=(u_1, u_2, \ldots, u_n)$.

Chapter 0

Overview

As the title suggests, this book presents several methods of nonlinear analysis for the treatment of nonlinear integral equations. To this end the book is developed on two levels which interfere. The first level is devoted to the abstract results of nonlinear analysis: compactness criteria, fixed point theorems, critical point results, and general principles of iterative approximation. The second level is devoted to the applications for systems of nonlinear integral equations (nonlinear integral equations in $\mathbf{R}^n$). Here we present existence, uniqueness, localization, and approximation results for Fredholm, Volterra, and Hammerstein integral equations, and an integral equation with delay.

Although most of the methods are classical, in several cases new points of view, extensions and new applications are presented. Thus the selected topics include: a unified existence theory for continuous and integrable solutions of integral equations; Schechter's bounded mountain pass theorem; a nontrivial solvability theory for Hammerstein integral equations; a localization result for a superlinear elliptic problem; the discrete continuation principle for generalized contractions; new fixed point results for increasing and decreasing operators in ordered Banach spaces; monotone iterative techniques; and the quasilinearization method for a delay integral equation from biomathematics. Of course, the selective topics reflect the particular interests of the author.

Let us now briefly describe the contents of this book without insisting on the exact meaning of the notation and notions which are used.

Chapter 1 presents basic concepts and results of compactness in general metric spaces, and in particular, in the spaces $C(K;\mathbf{R}^n)$ and $L^p(\Omega;\mathbf{R}^n)$, where K is a compact metric space, $\Omega \subset \mathbf{R}^N$ is bounded open and

$1 \le p < \infty$. We state and prove Hausdorff's theorem of characterization of the relatively compact subsets of a complete metric space, in terms of finite or relatively compact ε-nets. Then we prove the Ascoli–Arzèla compactness criterion in $C(K;\mathbf{R}^n)$, and the Fréchet–Kolmogorov theorem of characterization of relatively compact subsets of $L^p(\Omega;\mathbf{R}^n)$ $(1 \le p < \infty)$. We conclude this chapter by the following unified compactness criterion for both $C(\overline{\Omega};\mathbf{R}^n)$ and $L^p(\Omega;\mathbf{R}^n)$.

Theorem 0.1 *Let $\Omega \subset \mathbf{R}^N$ be bounded open and $p \in [1,\infty]$. Let $Y \subset L^p(\Omega;\mathbf{R}^n)$ for $1 \le p < \infty$, and $Y \subset C(\overline{\Omega};\mathbf{R}^n)$ for $p = \infty$. Assume that there exists a function $\nu \in L^p(\Omega)$ such that*

$$|u(x)| \le \nu(x)$$

for almost every $x \in \Omega$ and all $u \in Y$. Then Y is relatively compact in $L^p(\Omega;\mathbf{R}^n)$ if and only if

$$\sup_{u\in Y} |\tau_h(u) - u|_{L^p(\Omega_h;\mathbf{R}^n)} \to 0 \quad as \ h \to 0.$$

Here $\Omega_h = \Omega \cap (\Omega - h)$ and $\tau_h(u)(x) = u(x+h)$ for all $x \in \Omega_h$.

Chapter 2 is devoted to the concept of a completely continuous operator. We prove the theorem of representation of the completely continuous operators as uniform limits of sequences of continuous operators of finite rank. Also we present the proof of one of the fundamental results of nonlinear analysis, namely Schauder's fixed point theorem:

Theorem 0.2 *Let X be a Banach space, $D \subset X$ a nonempty convex bounded closed set and let $T : D \to D$ be a completely continuous operator. Then T has at least one fixed point.*

In Chapter 3 we present three examples of completely continuous operators arising from the theory of integral equations: the Fredholm integral operator $T : C(\overline{\Omega};\mathbf{R}^n) \to C(\overline{\Omega};\mathbf{R}^n)$ given by

$$T(u)(x) = \int_\Omega h(x,y,u(y))\,dy \quad (x \in \overline{\Omega}),$$

where $\Omega \subset \mathbf{R}^N$ is bounded open and $h : \overline{\Omega}^2 \times \mathbf{R}^n \to \mathbf{R}^n$ is continuous; the Volterra integral operator $T : C([a,b];\mathbf{R}^n) \to C([a,b];\mathbf{R}^n)$,

$$T(u)(t) = \int_a^t h(t,s,u(s))\,ds \quad (t \in [a,b]),$$

with $h : [a,b]^2 \times \mathbf{R}^n \to \mathbf{R}^n$ continuous, and the delay integral operator $T : D(T) \subset C([0,t_1];\mathbf{R}^n) \to C([0,t_1];\mathbf{R}^n)$ defined by

$$T(u)(t) = \int_{t-\tau}^{t} f(s, \widetilde{u}(s))\,ds \quad (t \in [0,t_1]).$$

Here $\tau > 0$ and $\varphi \in C([-\tau,0];\mathbf{R}^n)$ are given, $D(T) = \{u \in C([0,t_1];\mathbf{R}^n) : u(0) = \varphi(0)\}$, $\widetilde{u}(t) = \varphi(t)$ for $t \in [-\tau,0]$, $\widetilde{u}(t) = u(t)$ for $t \in [0,t_1]$, and $f \in C([-\tau,t_1] \times \mathbf{R}^n;\mathbf{R}^n)$ satisfies

$$\varphi(0) = \int_{-\tau}^{0} f(s,\varphi(s))\,ds.$$

The complete continuity of these operators is proved via Ascoli–Arzèla theorem (equivalently, via Theorem 0.1) and is used in order to deduce from Theorem 0.2, the existence of continuous solutions to the corresponding integral equation $u = T(u)$.

Chapter 4 is devoted to the Leray–Schauder principle and its applications to integral equations. In Section 4.1 we present an elementary proof (based upon Urysohn's lemma and Schauder's fixed point theorem) of the following version of the Leray–Schauder principle:

Theorem 0.3 *Let X be a Banach space, $K \subset X$ a closed convex subset, $U \subset K$ a bounded set, open in K and $u_0 \in U$ a fixed element. Assume that the operator $T : \overline{U} \to K$ is completely continuous and satisfies the boundary condition*

$$u \neq (1-\lambda)\,u_0 + \lambda T(u) \quad \textit{for all } u \in \partial U,\ \lambda \in (0,1).$$

Then T has at least one fixed point in $\overline{U}$.

Using Theorem 0.3 in Section 4.2 we establish an existence principle for the Fredholm integral equation which, in particular, yields existence results for Hammerstein integral equations and two-point boundary value problems for second order differential equations. Similar results for Volterra integral equations and, in particular, for Volterra–Hammerstein equations and the Cauchy problem for a first order differential system are established in Section 4.3. In Sections 4.4 and 4.5 we focus on the delay integral equation

$$u(t) = \int_{t-\tau}^{t} f(s, u(s))\,ds \tag{0.1}$$

which comes from biomathematics. We discuss the existence of solutions to the initial value problem and of periodic solutions.

In Chapter 5 we first examine the Nemytskii superposition operator, the Fredholm linear integral operator and the Hammerstein integral operator as mappings acting in L^p spaces. Then using again the Leray–Schauder principle we present an existence theory for the Hammerstein integral equation in $\mathbf{R}^n$

$$u(x) = \int_\Omega \kappa(x,y)\, f(y,u(y))\, dy \quad \text{a.e. on } \Omega, \tag{0.2}$$

in a such way that the two cases of continuous solutions and of L^p solutions $(1 \leq p < \infty)$ are treated together by considering $1 \leq p \leq \infty$. The main result is the following existence principle:

Theorem 0.4 *Let $\Omega \subset \mathbf{R}^N$ be a bounded open set, $\kappa : \Omega^2 \to \mathbf{R}$ and $f : \Omega \times \mathbf{R}^n \to \mathbf{R}^n$. Assume that there exists $p \in [1,\infty]$ and $q \in [1,\infty)$ such that the following conditions are satisfied:*

$$\begin{cases} \text{if } 1 \leq p < \infty \text{ then } \kappa \in L^p(\Omega; L^r(\Omega)); \\ \text{if } p = \infty \text{ then } \kappa \in C(\overline{\Omega}; L^r(\Omega)) \end{cases}$$

(here $1/q + 1/r = 1$) and

$$\begin{cases} \text{if } 1 \leq p < \infty \text{ then } |f(x,z)| \leq g(x) + c\,|z|^{p/q} \text{ for a.e. } x \in \Omega, \\ \text{all } z \in \mathbf{R}^n \text{ and some } g \in L^q(\Omega; \mathbf{R}_+),\ c \in \mathbf{R}_+; \\ \text{if } p = \infty \text{ then for every } R > 0 \text{ there is a } g_R \in L^q(\Omega) \text{ with} \\ |f(x,z)| \leq g_R(x) \text{ for a.e. } x \in \Omega \text{ and all } z \in \mathbf{R}^n \text{ with } |z| \leq R. \end{cases}$$

In addition assume the existence of a bounded open set $U \subset L^p(\Omega; \mathbf{R}^n)$ containing the null function such that

$$\begin{cases} \text{if } u \in \overline{U},\ \lambda \in (0,1) \text{ and} \\ u(x) = \lambda \int_\Omega \kappa(x,y)\, f(y,u(y))\, dy \text{ a.e. on } \Omega, \\ \text{then } u \in U. \end{cases} \tag{0.3}$$

Then (0.2) has at least one solution u in $L^p(\Omega; \mathbf{R}^n)$ with $u \in \overline{U}$. Moreover, for $p = \infty$, $u \in C(\overline{\Omega}; \mathbf{R}^n)$.

In Section 5.5 we give sufficient conditions so that (0.3) holds in the case of the Volterra–Hammerstein equation in $\mathbf{R}^n$

$$u(t) = \int_a^t \kappa(t,s)\, f(s,u(s))\, ds \quad \text{a.e. on } (a,b).$$

In particular, we discuss the existence of global solutions for the Cauchy problem

$$\begin{cases} u'(t) = f(t, u(t)) & \text{a.e. on } (a, b), \\ u(a) = 0, \end{cases}$$

where $f : [a, b] \times \mathbf{R}^n \to \mathbf{R}^n$.

In the first two sections of Chapter 6 we review basic properties of self-adjoint linear operators in Hilbert spaces. These are used in Section 6.3 where we present the Krasnoselskii's result concerning the splitting of a bounded linear operator $A : L^q(\Omega; \mathbf{R}^n) \to L^p(\Omega; \mathbf{R}^n)$ $(1/p + 1/q = 1)$ in the form

$$A = HH^*,$$

where $H : L^2(\Omega; \mathbf{R}^n) \to L^p(\Omega; \mathbf{R}^n)$ and H^* is the adjoint of H. Clearly solving (0.2) in $L^p(\Omega; \mathbf{R}^n)$ is equivalent to solving the operator equation

$$u = AN_f(u), \quad u \in L^p(\Omega; \mathbf{R}^n), \tag{0.4}$$

where

$$N_f : L^p(\Omega; \mathbf{R}^n) \to L^q(\Omega; \mathbf{R}^n), \ N_f(u)(x) = f(x, u(x))$$

and

$$A : L^q(\Omega; \mathbf{R}^n) \to L^p(\Omega; \mathbf{R}^n), \ A(u)(x) = \int_\Omega \kappa(x, y) u(y)\, dy.$$

We finish Section 6.3 by showing that if A splits as above, then we may replace the operator equation (0.4) by the equation

$$v = H^* N_f H(v), \quad v \in L^2(\Omega; \mathbf{R}^n) \tag{0.5}$$

in the Hilbert space $L^2(\Omega; \mathbf{R}^n)$. More exactly we have:

Proposition 0.1 *If* $u \in L^p(\Omega; \mathbf{R}^n)$ *solves* (0.4), *then* $v = H^* N_f(u)$ *is a solution of* (0.5). *Conversely, if* $v \in L^2(\Omega; \mathbf{R}^n)$ *solves* (0.5), *then* $u = H(v)$ *is a solution of* (0.4). *Moreover, there is a one-to-one correspondence between the solutions of* (0.4) *and the solutions of* (0.5).

In Chapter 7 we build the Golomb functional $E : L^2(\Omega; \mathbf{R}^n) \to \mathbf{R}$,

$$E(v) = \frac{1}{2}|v|_2^2 - \int_\Omega F(x, H(v)(x))\, dx,$$

where F is the potential of f. The Fréchet derivative E' of E is the operator

$$I - H^* N_f H : L^2(\Omega; \mathbf{R}^n) \to L^2(\Omega; \mathbf{R}^n),$$

and so the equation (0.5) can be written in the variational form

$$E'(v) = 0, \quad v \in L^2(\Omega; \mathbf{R}^n),$$

and its solutions appear as critical points of E. In Section 7.2 we prove the infinite-dimensional version of the classical Fermat's theorem about the connection between extremum points and critical points, and we give sufficient conditions for that a functional admits minimizers. In Section 7.3 the abstract results are applied to establish the existence of L^p solutions to (0.4).

Chapter 8 is devoted to the mountain pass principle. We first present the Ambrosetti–Rabinowitz mountain pass theorem and its proof based on Ekeland's variational principle. Then we state and prove Schechter's bounded mountain pass theorem which guarantees the existence of a critical point in a given ball of the space:

Theorem 0.5 *Let X be a Hilbert space, $R \in (0, \infty]$, $B_R = \{u \in X : |u| < R\}$ and $E \in C^1(\overline{B}_R)$. Assume that for some $\nu_0 > 0$, $(E'(u), u) \geq -\nu_0$ for all $u \in \partial B_R$, and that there are $u_0, u_1 \in \overline{B}_R$ and r with $|u_0| < r < |u_1|$ such that*

$$\max\{E(u_0), E(u_1)\} < \inf\{E(u) : u \in \overline{B}_R, \ |u| = r\}.$$

Let $\Gamma_R = \{\gamma \in C([0,1]; \overline{B}_R) : \gamma(0) = u_0, \ \gamma(1) = u_1\}$ and

$$c_R = \inf_{\gamma \in \Gamma_R} \max_{t \in [0,1]} E(\gamma(t)).$$

Then either there is a sequence of elements $u_k \in \overline{B}_R$ with

$$E(u_k) \to c_R, \quad E'(u_k) \to 0,$$

or there is a sequence of elements $u_k \in \partial B_R$ such that

$$E(u_k) \to c_R, \ \ E'(u_k) - \frac{(E'(u_k), u_k)}{R^2} u_k \to 0, \ \ (E'(u_k), u_k) \leq 0.$$

If in addition E satisfies the $(PS)_R$ condition and

$$E'(u) + \mu u \neq 0, \ \ u \in \partial B_R, \ \mu > 0, \tag{0.6}$$

then there exists an element $u \in \overline{B}_R \setminus \{u_0, u_1\}$ with

$$E(u) = c_R, \quad E'(u) = 0.$$

We point out that for $R<\infty$, (0.6) is the Leray–Schauder boundary condition for the operator $I-E'$, i.e., it is equivalent to

$$u \neq \lambda\left(I-E'\right)(u), \quad u \in \partial B_R,\ \lambda \in (0,1).$$

In Chapter 9 we apply Schechter's theorem to establish the nontrivial solvability of the abstract operator equation of Hammerstein type

$$u = AN(u), \quad u \in Y, \tag{0.7}$$

where Y is a Banach space, $N: Y \to Y^*$, and $A: Y^* \to Y$ is a linear operator. In Section 9.1 we establish the following existence and localization principle:

Theorem 0.6 *Assume that A splits into $A = HH^*$, where $H: X \to Y$ is a bounded linear operator, X is a Hilbert space, $H^*: Y^* \to X$ is the adjoint of H, and $N = J'$ for some functional $J \in C^1(Y;\mathbf{R})$ satisfying $J(0)=0$. In addition assume that $N(0)=0$ and the functional $(N(.),.)$ sends bounded sets into upper bounded sets and that there are $v_1 \in X\setminus\{0\}$, $r \in (0,|v_1|)$ and $R \geq |v_1|$ such that the following conditions are satisfied:*

$$\max\left\{0, \frac{1}{2}|v_1|_X^2 - JH(v_1)\right\} < \inf\left\{\frac{1}{2}|v|_X^2 - JH(v) : |v|_X = r\right\},$$

$$v \neq \lambda H^* NH(v) \quad \textit{for } |v|_X = R,\ \lambda \in (0,1),$$

E satisfies the $(PS)_R$ condition.

Then there exists a $v \in X\setminus\{0\}$ with $|v|_X \leq R$ such that $u = H(v)$ is a non-zero solution of (0.7).

We note that in our setting the equation (0.7) in Y can be converted into the equation $v = H^*NH(v)$ in X, which is equivalent to the critical point problem $E'(v) = 0$ for the energy functional

$$E: X \to \mathbf{R}, \quad E(v) = \frac{1}{2}|v|_X^2 - JH(v).$$

In Section 9.2 Theorem 0.6 is specialized for Hammerstein integral equations. Finally, in Section 9.3 we establish a localization result in a shell of $L^2(\Omega)$ for the superlinear Dirichlet problem

$$\begin{cases} -\Delta u = |u|^{p-2}u & \text{in } \Omega, \\ u = 0 & \text{on } \partial\Omega. \end{cases} \tag{0.8}$$

More exactly, we prove:

Theorem 0.7 *Let $\Omega \subset \mathbf{R}^N$ be a bounded open set with C^2-boundary and let $p \in (2, 2N/(N-2))$ if $N \geq 3$, and $p \in (2, \infty)$ if $N = 1$ or $N = 2$. Then the problem* (0.8) *has a solution u with*

$$|H|^{-1}\left[(p-1)\lambda_{p-1}\right]^{1/(p-2)} \leq \left|H^{-1}(u)\right|_2 \leq |H|^{-p/(p-2)}.$$

Here H is the square root of the operator $(-\Delta)^{-1}$ considered on $L^2(\Omega)$, $|H| = \sup\{|H(v)|_p : v \in L^2(\Omega),\ |v|_2 = 1\}$ and

$$\lambda_{p-1} = \inf\left\{\frac{\int_\Omega |u|^{p-2}|\nabla u|^2\,dx}{\left(\int_\Omega |u|^{p^2/2}\,dx\right)^{2/p}} : u \in C^1(\overline{\Omega}) \setminus \{0\},\ u = 0 \text{ on } \partial\Omega\right\}.$$

Chapter 10 presents the discrete continuation principle for contractions on spaces endowed with vector-valued metrics and one application to Hammerstein integral equations in $\mathbf{R}^n$ with matrix kernels.

Chapter 11 is devoted to the monotone iterative methods. The basic notion in this chapter is that of an ordered Banach space. We try to localize solutions of an operator equation in an ordered Banach space X, say $u = T(u)$, in an order interval $[u_0, v_0]$. In addition we look for solutions which are limits of increasing or decreasing sequences of elements of X. The basic property of the operator T is the monotonicity. This, combined with certain properties of the ordered Banach space X, guarantees the convergence of monotone sequences. Thus this chapter explores the contribution of the monotonicity to compactness. In Section 11.1 we review some notions and properties of the theory of ordered Banach spaces, paying special attention to the conditions which guarantee the convergence of bounded monotone sequences. Section 11.2 presents some fixed point theorems for monotone operators in general ordered Banach spaces. The method is then used in Sections 11.3, 11.4 to discuss the existence, uniqueness, localization, and monotone iterative approximation of solutions for the Fredholm integral equation and the delay integral equation already considered in Chapters 3 and 4. In Section 11.5 monotone iterative methods are used to localize and approximate solutions of the abstract Hammerstein equation in $\mathbf{R}^n$

$$u(x) = AN_f(u)(x) \quad \text{a.e. on } \Omega. \tag{0.9}$$

Here N_f is Nemytskii's superposition operator associated to a given function $f : \Omega \times \mathbf{R}^n \to \mathbf{R}^n$ ($\Omega \subset \mathbf{R}^N$ bounded open), and A is a bounded linear operator from $L^q(\Omega; \mathbf{R}^n)$ to $L^p(\Omega; \mathbf{R})$. We seek solutions u in an order interval $[u_0, v_0]$ of $L^p(\Omega; \mathbf{R}^n)$, where u_0 is a lower solution of (0.9) and v_0 is an upper solution of (0.9). Of course, we assume that N_f maps $[u_0, v_0]$ into $L^q(\Omega; \mathbf{R}^n)$. The main assumption is the monotonicity of f in its second argument. The advantage of working in the space $L^p(\Omega; \mathbf{R}^n)$ $(1 \leq p < \infty)$ ordered by the regular cone of all positive functions is that we do not need the complete continuity of the operator $T = AN_f$. Additional properties of the operator A in connexion with its spectrum, allow us to build upper and lower solutions for (0.9) and, consequently, to obtain extremal solutions of (0.9).

Finally, Chapter 12 discusses the problem of the convergence order of the iterative methods. After explaining that the convergence of the methods described in Chapters 10 and 11 is at most linear, this chapter briefly presents Newton's method and its version for differential equations, the quasilinearization method. The last one is addapted for the quadratic bilateral monotone approximation of the solution to the initial value problem for the delay integral equation (0.1).

Part I

FIXED POINT METHODS

Chapter 1

Compactness in Metric Spaces

In this chapter we first define the notions of a compact metric space and of a relatively compact subset of a metric space. Then we state and prove Hausdorff's theorem of the characterization of the relatively compact subsets of a complete metric space in terms of finite and relatively compact ε-nets. Furthermore, we prove the Ascoli–Arzèla and Fréchet–Kolmogorov theorems of characterization of the relatively compact subsets of $C(K;\mathbf{R}^n)$ and $L^p(\Omega;\mathbf{R}^n)$, respectively. Here K is a compact metric space, $\Omega \subset \mathbf{R}^N$ is a bounded open set and $1 \leq p < \infty$.

1.1 Hausdorff's Theorem

Proposition 1.1 *Let (X,d) be a metric space. The following statements are equivalent:*

(a) Every sequence of elements of X has a convergent subsequence in X.

(b) The space X is complete and for each $\varepsilon > 0$ it admits a finite covering by open balls of radius ε.

Proof. (a) $\Rightarrow$ (b) : The completeness of X follows from (a) since every Cauchy sequence which has a convergent subsequence is convergent. Assume that for some $\varepsilon > 0$ the space X does not admit a finite covering by balls of radius ε. Then, for any fixed element $u_1 \in X$ there exists $u_2 \in X$ with $d(u_1,u_2) \geq \varepsilon$. Furthermore, there exists $u_3 \in X$ such that $d(u_1,u_3) \geq \varepsilon$ and $d(u_2,u_3) \geq \varepsilon$, and so on. Thus we find a sequence (u_k) of elements of

X such that

$$d(u_k, u_j) \geq \varepsilon$$

for $k \neq j$. This shows that (u_k) has no convergent subsequences, which contradicts (a).

(b) $\Rightarrow$ (a) : Let (u_k) be an arbitrary sequence of elements of X and let (ε_k) be any decreasing sequence of positive numbers converging to zero. According to (b) there exists $v_1 \in X$ such that the open ball $B_{\varepsilon_1}(v_1)$ (of center v_1 and radius ε_1) contains infinitely many terms of the sequence (u_k). Similarly there exists $v_2 \in X$ such that $B_{\varepsilon_2}(v_2)$ contains infinitely many terms from those contained by $B_{\varepsilon_1}(v_1)$, and so on. Thus we find a subsequence (u_{k_j}) of (u_k) such that $u_{k_1} \in B_{\varepsilon_1}(v_1)$, $u_{k_2} \in B_{\varepsilon_1}(v_1) \cap B_{\varepsilon_2}(v_2)$ and generally

$$u_{k_j} \in \bigcap_{i=1}^{j} B_{\varepsilon_i}(v_i)$$

for $j = 1, 2, \ldots$. Since $u_{k_i}, u_{k_j} \in B_{\varepsilon_i}(v_i)$ for $i < j$, we have

$$d(u_{k_i}, u_{k_j}) \leq d(u_{k_i}, v_i) + d(v_i, u_{k_j}) < 2\varepsilon_i.$$

It follows that (u_{k_j}) is a Cauchy sequence. Since X is complete this subsequence is convergent. ∎

Definition 1.1 A metric space X is said to be *compact* if it satisfies condition (a) (equivalently (b)) in Proposition 1.1.

A subset Y of a metric space X is said to be *relatively compact* if its closure $\overline{Y}$ is compact (as a metric subspace of X).

Definition 1.2 Let (X, d) be a metric space, Y a subset of X and $\varepsilon > 0$. A subset $R \subset X$ is said to be an *ε-net* for Y if for every $u \in Y$ there exists a $v \in R$ such that $d(u, v) < \varepsilon$.

Notice that a metric space is compact if and only if it is complete and for every $\varepsilon > 0$ it admits a finite ε-net. Consequently every compact metric space is complete and bounded (in the sense that its metric takes values in a bounded real interval).

Theorem 1.1 (Hausdorff) *Let (X, d) be a complete metric space and $Y \subset X$ be a subset. The following statements are equivalent:*

(a) Y is relatively compact.
(b) For every $\varepsilon > 0$ there exists in X a finite ε-net for Y.
(c) For every $\varepsilon > 0$ there exists in X a relatively compact ε-net for Y.

Proof. (a)$\Rightarrow$(b): Assume that Y is relatively compact. Then $\overline{Y}$ is compact, and so for every $\varepsilon > 0$ there exists in $\overline{Y}$ a finite ε-net $\mathcal{R}$ for $\overline{Y}$. Clearly, $\mathcal{R}$ is a finite ε-net (in X) for Y.

(b)$\Rightarrow$(a): Assume (b) holds. Then for a given $\varepsilon > 0$ there exists in X a finite $\varepsilon/3$-net $\mathcal{R}$ for Y. For every $u \in \mathcal{R}$ we choose an element $v_u \in Y$ such that $d(u, v_u) < \varepsilon/3$ (if such an element exists). It is easy to see that the set

$$\mathcal{R}' := \{v_u : u \in \mathcal{R}\}$$

is a finite $2\varepsilon/3$-net for Y and also an ε-net for $\overline{Y}$. Hence $\overline{Y}$ is compact.

(b)$\Rightarrow$(c): This is trivial, since every finite set is compact.

(c)$\Rightarrow$(b): Assume (c) holds and take any $\varepsilon > 0$. Then there exists in X a relatively compact $\varepsilon/2$-net $\mathcal{R}$ for Y. Furthermore, for the compact set $\overline{\mathcal{R}}$ there exists a finite $\varepsilon/2$-net $\mathcal{R}'$. It is easy to see that $\mathcal{R}'$ is an ε-net (in X) for Y. Hence (b) is true. ■

We conclude this section by the following compactness property of the uniform limit of a family of relatively compact subsets of a complete metric space.

Proposition 1.2 *Let (X, d) be a complete metric space and let $\{Y_\lambda : \lambda > 0\}$ be a family of relatively compact subsets of X. Assume that $Y \subset X$ is the uniform limit of Y_λ as $\lambda \to 0$, that is for each $\varepsilon > 0$ there is $\lambda_\varepsilon > 0$ such that for every $u \in Y$ and $\lambda \in (0, \lambda_\varepsilon)$ there exists $u_\lambda \in Y_\lambda$ with $d(u, u_\lambda) < \varepsilon$. Then Y is relatively compact.*

The proof can be immediately given in terms of finite ε-nets. We leave the details to the reader.

In the next two sections, as applications of Hausdorff's theorem, we shall prove the Ascoli–Arzèla and Fréchet–Kolmogorov theorems of characterization of the relatively compact subsets in $C(K; \mathbf{R}^n)$ and $L^p(\Omega; \mathbf{R}^n)$, respectively.

1.2 The Ascoli–Arzèla Theorem

Let (K, d) be a compact metric space and $C(K; \mathbf{R}^n)$ be the Banach space of all continuous functions from K to $\mathbf{R}^n$, under the sup-norm $|\cdot|_\infty$.

Theorem 1.2 (Ascoli–Arzèla) *A subset Y of $C(K; \mathbf{R}^n)$ is relatively compact if and only if the following conditions are satisfied:*

(i) Y is bounded, i.e., there exists a constant $c > 0$ such that

$$|u(x)| \le c$$

for all $x \in K$ and $u \in Y$.

(ii) Y is equicontinuous, i.e., for every $\varepsilon > 0$ there exists a $\delta > 0$ such that for all $u \in Y$,

$$|u(x) - u(x')| < \varepsilon$$

whenever $x,\ x' \in K$ and $d(x, x') < \delta$.

Proof. Assume that Y is relatively compact in $C(K; \mathbf{R}^n)$. Then, obviously, it is bounded and according to Hausdorff's theorem, for every $\varepsilon > 0$ there exists in $C(K; \mathbf{R}^n)$ a finite $\varepsilon/3$-net $\mathcal{R}$ for Y. Let

$$\mathcal{R} = \{u_1, u_2, \ldots, u_m\}.$$

The functions u_k being uniformly continuous on the compact K, there exists a $\delta > 0$ such that for all $k \in \{1, 2, \ldots, m\}$,

$$|u_k(x) - u_k(x')| < \varepsilon/3$$

whenever $x,\ x' \in K$ and $d(x, x') < \delta$. Now for any $u \in Y$ there exists a $k \in \{1, 2, \ldots, m\}$ with

$$|u - u_k|_\infty < \varepsilon/3,$$

so if $d(x, x') < \delta$, we have

$$\begin{aligned} & |u(x) - u(x')| \\ \le\ & |u(x) - u_k(x)| + |u_k(x) - u_k(x')| + |u_k(x') - u(x')| \\ <\ & \varepsilon/3 + \varepsilon/3 + \varepsilon/3 = \varepsilon. \end{aligned}$$

This shows that Y is equicontinuous.

Conversely, assume that (i) and (ii) hold. The space $C(K; \mathbf{R}^n)$ being a closed subspace of the Banach space $B(K; \mathbf{R}^n)$ of all bounded functions from K to $\mathbf{R}^n$, it is sufficient to prove that Y is relatively compact in $B(K; \mathbf{R}^n)$. For this, according to Hausdorff's theorem, we have to show that for every $\varepsilon > 0$ there exists in $B(K; \mathbf{R}^n)$ a relatively compact ε-net for Y. For a fixed $\varepsilon > 0$ we consider $\delta > 0$ given by condition (ii). Since

K is compact, there is a finite $\delta/2$-net $\{x_1, x_2, ..., x_m\}$ for K. Let

$$\begin{aligned} K_1 &:= B_{\delta/2}(x_1), \\ K_2 &:= B_{\delta/2}(x_2) \setminus K_1, \\ &\vdots \\ K_m &:= B_{\delta/2}(x_m) \setminus \bigcup_{j=1}^{m-1} K_j. \end{aligned}$$

One has

$$K = \bigcup_{j=1}^{m} K_j, \quad K_i \cap K_j = \emptyset \quad \text{for} \quad i \neq j,$$

and

$$d(x, x') < \delta \quad \text{for all} \quad x,\ x' \in K_j, \quad j = 1,\ 2,\ ...,\ m.$$

Let φ_j be the characteristic function of K_j, i.e.,

$$\varphi_j(x) = \begin{cases} 1 & \text{for } x \in K_j \\ 0 & \text{for } x \in K \setminus K_j. \end{cases}$$

We consider the following set of bounded functions from K to $\mathbf{R}^n$,

$$\mathcal{R} = \left\{ \sum_{j=1}^{m} \varphi_j \lambda_j : \lambda_j \in \mathbf{R}^n,\ |\lambda_j| \leq c,\ j = 1, 2, ..., m \right\}.$$

Here c is the constant in (i). It is easy to see that $\mathcal{R}$ is relatively compact in $B(K; \mathbf{R}^n)$ since the convergence of a sequence of functions in $\mathcal{R}$ reduces to the convergence of the corresponding sequences of vectors λ_j. Thus it remains to prove that $\mathcal{R}$ is an ε-net for Y. For this let us choose an element $\overline{x}_j \in K_j$ for every $j \in \{1,\ 2,\ ...,\ m\}$. For each $u \in Y$ we build the function

$$v(x) = \sum_{j=1}^{m} \varphi_j(x)\, u(\overline{x}_j) \quad (x \in K).$$

Obviously $v \in \mathcal{R}$ since $|u(\overline{x}_j)| \leq c$. In addition, for any $x \in K$ there exists a j with $x \in K_j$, and so $v(x) = u(\overline{x}_j)$. Hence by (ii)

$$|u(x) - v(x)| = |u(x) - u(\overline{x}_j)| < \varepsilon$$

since $d(x, \overline{x}_j) < \delta$. Consequently $|u - v|_\infty < \varepsilon$, which shows that $\mathcal{R}$ is an ε-net for Y. ■

Let $|.|_{k,\infty}$ denote the norm of $C^k(\overline{\Omega}; \mathbf{R}^n)$ ($k \in \mathbf{N} \setminus \{0\}$, $\Omega \subset \mathbf{R}^N$ bounded open),

$$|u|_{k,\infty} = \max\left\{|u|_\infty, |u'|_\infty, ..., \left|u^{(k)}\right|_\infty\right\}.$$

Corollary 1.1 *Let Ω be a bounded open subset of $\mathbf{R}^N$. Every bounded subset of the space $(C^1(\overline{\Omega}; \mathbf{R}^n), |.|_{1,\infty})$ is relatively compact in $(C(\overline{\Omega}; \mathbf{R}^n), |.|_\infty)$.*

Proof. Let Y be a bounded subset of $\left(C^1(\overline{\Omega}; \mathbf{R}^n), |.|_{1,\infty}\right)$. Then Y is also bounded in $(C(\overline{\Omega}; \mathbf{R}^n), |.|_\infty)$. In addition, all functions in Y are Lipschitz with the same Lipschitz constant, and therefore Y is equicontinuous. Now the Ascoli–Arzèla theorem guarantees that Y is relatively compact in $(C(\overline{\Omega}; \mathbf{R}^n), |.|_\infty)$. ■

Remark 1.1 Let Ω be a bounded open subset of R^N and $k \in N \setminus \{0\}$. Every bounded subset of the space $\left(C^k(\overline{\Omega}; \mathbf{R}^n), |.|_{k,\infty}\right)$ is relatively compact in $\left(C^{k-1}(\overline{\Omega}; \mathbf{R}^n), |.|_{k-1,\infty}\right)$.

1.3 The Fréchet–Kolmogorov Theorem

In this section we present some basic properties of the spaces $L^p(\Omega; \mathbf{R}^n)$ and the L^p version of the Ascoli–Arzèla theorem.

Let $\Omega \subset \mathbf{R}^N$ be open and $p \in \mathbf{R}$, $1 \le p < \infty$. Let $L^p(\Omega; \mathbf{R}^n)$ be the space of all measurable functions $u : \Omega \to \mathbf{R}^n$ such that $|u|^p$ is Lebesgue integrable on Ω. The space $L^p(\Omega; \mathbf{R}^n)$ is endowed with the norm

$$|u|_p = \left(\int_\Omega |u(x)|^p \, dx\right)^{1/p}.$$

For $p = \infty$ we let $L^\infty(\Omega; \mathbf{R}^n)$ be the space of all measurable functions $u : \Omega \to \mathbf{R}^n$ for which there exists a constant c with $|u(x)| \le c$ for almost every (a.e. for short) $x \in \Omega$. The norm on $L^\infty(\Omega; \mathbf{R}^n)$ is given by $|u|_\infty = \sup \text{ess} |u(x)|$, i.e.,

$$|u|_\infty = \inf\{c : |u(x)| \le c \text{ for a.e. } x \in \Omega\}.$$

Under the norm $|.|_p$, $L^p(\Omega; \mathbf{R}^n)$ is a Banach space for each $p \in [1, \infty]$. The space $L^2(\Omega; \mathbf{R}^n)$ is a Hilbert space with the inner-product

$$(u, v)_2 = \int_\Omega (u(x), v(x)) \, dx.$$

In some places we shall denote the norm $|.|_p$ by $|.|_{L^p(\Omega;\mathbf{R}^n)}$ in order to avoid any confusion.

The space $L^p(\Omega;\mathbf{R})$ is denoted by $L^p(\Omega)$. Also, for $\Omega = (a,b) \subset \mathbf{R}$ we write $L^p(a,b;\mathbf{R}^n)$ instead of $L^p((a,b);\mathbf{R}^n)$ and $L^p(a,b)$ instead of $L^p((a,b))$.

For $1 \le p \le \infty$ we let p' be the conjugate exponent of p, i.e., $1/p + 1/p' = 1$. Recall the well known Hölder's inequality: if $u \in L^p(\Omega;\mathbf{R}^n)$ and $v \in L^{p'}(\Omega;\mathbf{R}^n)$ then $|u|\,|v| \in L^1(\Omega)$ and

$$\int_\Omega |u(x)|\,|v(x)|\,dx \le |u|_p\,|v|_{p'}.$$

This inequality implies that if Ω is bounded then for $1 \le p < q \le \infty$, the following inclusions hold

$$C(\overline{\Omega};\mathbf{R}^n) \subset L^q(\Omega;\mathbf{R}^n) \subset L^p(\Omega;\mathbf{R}^n).$$

We also recall that the dual of $L^p(\Omega;\mathbf{R}^n)$ $(1 \le p < \infty)$ is the space $L^{p'}(\Omega;\mathbf{R}^n)$, where $1/p + 1/p' = 1$, whilst the dual of $L^\infty(\Omega;\mathbf{R}^n)$ strictly includes $L^1(\Omega;\mathbf{R}^n)$.

The following density result will be useful: the space $C_0(\Omega;\mathbf{R}^n)$ of all continuous functions from $\overline{\Omega}$ to $\mathbf{R}^n$ with compact support in Ω is dense in $L^p(\Omega;\mathbf{R}^n)$ for every $p \in [1,\infty)$.

For details and further information on the L^p spaces, we refer the reader to Brezis [6].

For the rest of this section we assume that $\Omega \subset \mathbf{R}^N$ is a bounded open set. For a given number $r > 0$ and any locally integrable (integrable on each compact subset) function $u : \mathbf{R}^N \to \mathbf{R}^n$ we define the *average function* of u with respect to radius r, $m_r(u) : \mathbf{R}^N \to \mathbf{R}^n$, by

$$m_r(u)(x) = \frac{1}{\omega_r}\int_{B_r(x)} u(y)\,dy \quad \left(x \in \mathbf{R}^N\right). \tag{1.1}$$

Here $\omega_r = \mu(B_r(0))$, the measure of the ball of radius r of $\mathbf{R}^N$. For a function $u \in L^p(\Omega;\mathbf{R}^n)$ we define the average function $m_r(u)$ in the same way assuming that $u(y) = 0$ for $y \notin \Omega$.

Lemma 1.1 *For every $u \in L^p(\Omega;\mathbf{R}^n)$ $(1 \le p < \infty)$ and every $r > 0$, the function $m_r(u)$ is continuous on $\mathbf{R}^N$ and satisfies*

$$|m_r(u)(x)| \le \omega_r^{-1/p}\,|u|_{L^p(\Omega;\mathbf{R}^n)}, \quad x \in \mathbf{R}^N, \tag{1.2}$$

and

$$|m_r(u)|_{L^p(\Omega;\mathbf{R}^n)} \le |u|_{L^p(\Omega;\mathbf{R}^n)}. \tag{1.3}$$

Proof. We have

$$\begin{aligned} \left|m_r(u)(x) - m_r(u)(x')\right| &= \omega_r^{-1}\left|\int_{B_r(x)} u(y)\,dy - \int_{B_r(x')} u(y)\,dy\right| \\ &\leq \omega_r^{-1}\int_{B_r(x,x')} |u(y)|\,dy, \end{aligned}$$

where

$$B_r(x,x') = (B_r(x)\cup B_r(x'))\setminus B_r(x)\cap B_r(x').$$

Hölder's inequality applied to u and to the constant function 1 yields

$$\left|m_r(u)(x) - m_r(u)(x')\right| \leq \omega_r^{-1}\mu(B_r(x,x'))^{1/p'}|u|_{L^p(\Omega;\mathbf{R}^n)}.$$

Letting $x'\to x$ we deduce that

$$m_r(u)(x') \to m_r(u)(x)$$

since $\mu(B_r(x,x'))\to 0$. Hence $m_r(u)$ is continuous at any point $x\in\mathbf{R}^N$.

Notice that (1.2) immediately follows from (1.1) if we apply Hölder's inequality to the functions 1 and u.

Finally, (1.3) is obtained if we apply Hölder's inequality and we interchange the order of integration as show the following inequalities:

$$\begin{aligned} |m_r(u)|_p &= \left(\int_\Omega |m_r(u)(x)|^p\,dx\right)^{1/p} \\ &\leq \left(\omega_r^{-1}\int_\Omega\left(\int_{B_r(x)}|u(y)|^p\,dy\right)dx\right)^{1/p} \\ &= \omega_r^{-1/p}\left(\int_\Omega\int_{B_r(0)}|u(x+y)|^p\,dydx\right)^{1/p} \\ &= \omega_r^{-1/p}\left(\int_{B_r(0)}\int_\Omega|u(x+y)|^p\,dxdy\right)^{1/p} \\ &\leq |u|_p. \end{aligned}$$

This concludes the proof. ■

For any $h\in\mathbf{R}^N$ and any $u\in L^p(\Omega;\mathbf{R}^n)$ we define the *translation* of u by h, to be the function $\tau_h(u)$ from $\mathbf{R}^N$ to $\mathbf{R}^n$, given by

$$\tau_h(u)(x) = \begin{cases} u(x+h) & \text{if } x+h\in\Omega \\ 0 & \text{if } x+h\in\mathbf{R}^N\setminus\Omega. \end{cases}$$

Theorem 1.3 (Fréchet–Kolmogorov) *Let $\Omega \subset \mathbf{R}^N$ be bounded open and $1 \leq p < \infty$. A subset Y of $L^p(\Omega; \mathbf{R}^n)$ is relatively compact if and only if the following conditions are satisfied:*

(i) Y is bounded, i.e., there exists a $c > 0$ such that

$$|u|_p \leq c$$

for all $u \in Y$;

(ii) $\tau_h(u) \to u$ in $L^p(\Omega; \mathbf{R}^n)$ as $h \to 0$, uniformly for $u \in Y$, i.e.,

$$\sup_{u \in Y} |\tau_h(u) - u|_p \to 0 \quad as \ h \to 0.$$

Proof. Assume that Y is relatively compact in $L^p(\Omega; \mathbf{R}^n)$. Then clearly (i) holds. Also, for a given $\varepsilon > 0$ there exists in $L^p(\Omega; \mathbf{R}^n)$ a finite $\varepsilon/3$-net for Y. Since $C(\overline{\Omega}; \mathbf{R}^n)$ is dense in $L^p(\Omega; \mathbf{R}^n)$ (recall that Ω is bounded), we may assume that the elements of the net belong to $C(\overline{\Omega}; \mathbf{R}^n)$. Let these elements be u_j, $j = 1, 2, ..., m$. Since u_j is uniform continuous on the compact set $\overline{\Omega}$, there exists a $\delta > 0$ such that

$$\begin{aligned} |u_j(x) - \tau_h(u_j)(x)| &= |u_j(x) - u_j(x+h)| \\ &< \frac{\varepsilon}{(2 \cdot 3^p \mu(\Omega))^{1/p}} \end{aligned}$$

for all $h \in \mathbf{R}^N$ with $|h| < \delta$, $x \in \Omega_h = \Omega \cap (\Omega - h)$ and $j = 1, 2, ..., m$. It follows that

$$\int_{\Omega_h} |u_j(x) - \tau_h(u_j)(x)|^p \, dx < \frac{\varepsilon^p}{2 \cdot 3^p}.$$

In addition, by (i) we may assume that for each j we have

$$\int_{\Omega \setminus \Omega_h} |u_j(x)|^p \, dx \leq \frac{\varepsilon^p}{2 \cdot 3^p}.$$

Consequently

$$|\tau_h(u_j) - u_j|_p < \frac{\varepsilon}{3}$$

for $|h| < \delta$ and $j = 1, 2, ..., m$. Now, for a given $u \in Y$ there is a $j \in \{1, 2, ..., m\}$ with $|u - u_j|_p < \varepsilon/3$. On the other hand, we have

$$\tau_h(u) - u = \tau_h(u - u_j) - (u - u_j) + (\tau_h(u_j) - u_j).$$

As a result, for $|h| < \delta$ one has

$$|\tau_h(u) - u|_p \leq |\tau_h(u - u_j)|_p + |u - u_j|_p + |\tau_h(u_j) - u_j|_p < \varepsilon.$$

Thus (ii) holds.

Conversely, assume (i) and (ii) are satisfied. For any $u \in Y$ one has

$$\begin{aligned}
&\left|m_r(u)(x) - m_r(u)(x')\right| \\
= \; &\omega_r^{-1} \left| \int_{B_r(x)} u(y)\,dy - \int_{B_r(x')} u(y)\,dy \right| \\
\leq \; &\omega_r^{-1} \int_{B_r(x)} \left|u(y) - \tau_{x'-x}(u)(y)\right| dy \\
\leq \; &\omega_r^{-1/p} \left|u - \tau_{x'-x}(u)\right|_p .
\end{aligned}$$

This inequality together with (ii) shows that the set

$$m_r(Y) = \{m_r(u) : u \in Y\}$$

is equicontinuous. In addition, according to (1.2) and (i) the set $m_r(Y)$ is bounded in $C(\overline{\Omega}; \mathbf{R}^n)$. Now the Ascoli–Arzèla theorem guarantees that $m_r(Y)$ is a relatively compact subset of $C(\overline{\Omega}; \mathbf{R}^n)$, and so of $L^p(\Omega; \mathbf{R}^n)$. On the other hand, from

$$\begin{aligned}
m_r(u)(x) - u(x) &= \omega_r^{-1} \int_{B_r(x)} (u(y) - u(x))\,dy \\
&= \omega_r^{-1} \int_{B_r(0)} (\tau_y(u)(x) - u(x))\,dy
\end{aligned}$$

we deduce

$$\left|m_r(u) - u\right|_p \leq \sup_{|y| \leq r} \left|\tau_y(u) - u\right|_p .$$

This shows that Y is the uniform limit (in $L^p(\Omega; \mathbf{R}^n)$) of $m_r(Y)$ as $r \to 0$. Now Proposition 1.2 guarantees that Y is relatively compact in $L^p(\Omega; \mathbf{R}^n)$. ■

Notice we can slightly modify the Fréchet–Kolmogorov theorem in order that a unified criterion for compactness in both $C(\overline{\Omega}; \mathbf{R}^n)$ and $L^p(\Omega; \mathbf{R}^n)$ $(1 \leq p < \infty)$ be possible. More exactly, we have the following theorem.

Theorem 1.4 *Let $\Omega \subset \mathbf{R}^N$ be bounded open and $p \in [1, \infty]$. Let $Y \subset L^p(\Omega; \mathbf{R}^n)$ for $1 \leq p < \infty$ and $Y \subset C(\overline{\Omega}; \mathbf{R}^n)$ for $p = \infty$. Assume that there exists a function $\nu \in L^p(\Omega)$ such that*

$$|u(x)| \leq \nu(x) \tag{1.4}$$

for a.e. $x \in \Omega$ *and all* $u \in Y$. *Then* Y *is relatively compact in* $L^p(\Omega; \mathbf{R}^n)$ *if and only if*

$$\sup_{u \in Y} |\tau_h(u) - u|_{L^p(\Omega_h; \mathbf{R}^n)} \to 0 \quad as \;\; h \to 0. \tag{1.5}$$

Here $\Omega_h = \Omega \cap (\Omega - h)$.

Proof. 1) Assume $p = \infty$. Then (1.4) is equivalent to the boundedness of Y in $C(\overline{\Omega}; \mathbf{R}^n)$, whilst (1.5) means exactly the equicontinuity of Y.

2) Let $1 \leq p < \infty$. Then (1.4) implies that Y is bounded in $L^p(\Omega; \mathbf{R}^n)$. Hence condition (i) in Theorem 1.3 holds. Now observe that (1.4) and (1.5) guarantee condition (ii) in Theorem 1.3. Indeed, one has

$$|\tau_h(u) - u|^p_{L^p(\Omega; \mathbf{R}^n)} = |\tau_h(u) - u|^p_{L^p(\Omega_h; \mathbf{R}^n)} + |u|^p_{L^p(\Omega \setminus \overline{\Omega}_h; \mathbf{R}^n)}.$$

Then using (1.4) we obtain

$$|\tau_h(u) - u|^p_{L^p(\Omega; \mathbf{R}^n)} \leq |\tau_h(u) - u|^p_{L^p(\Omega_h; \mathbf{R}^n)} + |\nu|^p_{L^p(\Omega \setminus \overline{\Omega}_h)}.$$

This inequality together with (1.5) guarantees (ii) in Theorem 1.3, since

$$|\nu|^p_{L^p(\Omega \setminus \overline{\Omega}_h)} \to 0 \quad \text{as} \;\; h \to 0.$$

Thus Theorem 1.3 applies and the proof is complete. ■

Chapter 2

Completely Continuous Operators on Banach Spaces

In this chapter we define the notion of a completely continuous operator from a Banach space to another Banach space and we present some simple properties of the completely continuous operators. Next we prove Brouwer's fixed point theorem. Finally, we prove the famous Schauder fixed point theorem which, like Banach's contraction principle, represents a fundamental result in nonlinear analysis.

2.1 Completely Continuous Operators

Definition 2.1 Let X, Y be Banach spaces and $T : D \subset X \to Y$.

(a) The operator T is said to be *bounded* if it maps any bounded subset of D into a bounded subset of Y.

(b) The operator T is said to be *completely continuous* if it is continuous and maps any bounded subset of D into a relatively compact subset of Y.

(c) The operator T is said to be *of finite rank* if $T(D)$ lies in a finite-dimensional subspace of Y.

It is clear that a continuous operator $T : D \subset X \to Y$ is completely continuous if and only if for every bounded sequence (u_k) with $u_k \in D$, the sequence $(T(u_k))$ has a convergent subsequence.

Notice that any completely continuous operator is a bounded operator.

Theorem 2.1 (1) *If the operators $T_1, T_2 : D \subset X \to Y$ are bounded (completely continuous) then for every $\alpha, \beta \in \mathbf{R}$ the operator $\alpha T_1 + \beta T_2$ is bounded (respectively, completely continuous).*

(2) *Let* X, Y, Z *be Banach spaces and* T_1, T_2 *be two operators which are defined as follows:*

$$D_1 \xrightarrow{T_1} T_1(D_1) \subset D_2 \xrightarrow{T_2} Z, \quad D_1 \subset X, \; D_2 \subset Y.$$

If both operators T_1, T_2 *are bounded then the composite operator* T_2T_1 *is also bounded. If one of the operators* T_1, T_2 *is bounded continuous and the other one is completely continuous, then* T_2T_1 *is completely continuous.*

Proof. Both statements (1) and (2) are simple consequences of Definition 2.1. For (2) also use the fact that a continuous operator maps relatively compact sets into relatively compact sets. ■

Theorem 2.2 (1) *If the operators* $T_k : D \to Y$, $D \subset X$, $k = 1, 2, ...$ *are completely continuous and* $T : D \to Y$ *is such that*

$$T(u) = \lim_{k\to\infty} T_k(u) \tag{2.1}$$

uniformly on any bounded subset of D, *then* T *is completely continuous too.*

(2) *Let* $D \subset X$ *be a bounded closed set and* $T : D \to Y$ *a completely continuous operator. Then there exists a sequence of continuous operators* $T_k : D \to Y$ *of finite rank such that* (2.1) *holds, uniformly on* D, *and*

$$T_k(D) \subset \text{conv}\,(T(D))$$

for every k.

Proof. (1) We first note that T is continuous at any point $u_0 \in D$. This follows, using (2.1) and the continuity of T_k, from the following inequality

$$\begin{aligned} |T(u) - T(u_0)|_Y \;\le\; & |T(u) - T_k(u)|_Y + |T_k(u) - T_k(u_0)|_Y \\ & + |T_k(u_0) - T(u_0)|_Y \,. \end{aligned}$$

Let $M \subset D$ be a bounded set. Since T_k is completely continuous we have that $T_k(M)$ is relatively compact. In addition, from (2.1) we have that $T(M)$ is the uniform limit of $T_k(M)$ as $k \to \infty$. By Proposition 1.2, $T(M)$ is relatively compact. Hence T is completely continuous.

(2) Since T is completely continuous and D is bounded, the set $T(D)$ is relatively compact. Consequently for every $\varepsilon > 0$ there exist finitely many elements $v_j \in T(D)$, $j = 1, 2, ..., m_\varepsilon$, such that

$$\overline{T(D)} \subset \bigcup_{j=1}^{m_\varepsilon} B_\varepsilon(v_j)\,.$$

Let (φ_j) be the partition of unity with respect to this covering of $\overline{T(D)}$. Hence

$$\varphi_j \in C\left(\overline{T(D)};[0,1]\right), \quad \operatorname{supp}\varphi_j \subset \overline{B}_\varepsilon(v_j), \quad \sum_{j=1}^{m_\varepsilon} \varphi_j(v) = 1$$

for all $v \in \overline{T(D)}$. We define the operator $T_\varepsilon : D \to Y$ by

$$T_\varepsilon(u) = \sum_{j=1}^{m_\varepsilon} \varphi_j(T(u))\, v_j \quad (u \in D).$$

By its construction T_ε is a continuous operator of finite rank and

$$T_\varepsilon(D) \subset \operatorname{conv}(T(D)).$$

In addition, we have

$$\begin{aligned} |T(u) - T_\varepsilon(u)|_Y &= \left| \sum_{j=1}^{m_\varepsilon} \varphi_j(T(u))\,(v_j - T(u)) \right|_Y \\ &\le \sum_{j=1}^{m_\varepsilon} \varphi_j(T(u))\, |v_j - T(u)|_Y \\ &\le \varepsilon \sum_{j=1}^{m_\varepsilon} \varphi_j(T(u)) = \varepsilon \end{aligned}$$

since $T(u) \in B_\varepsilon(v_j)$ whenever $\varphi_j(T(u)) > 0$. Hence

$$T(u) = \lim_{\varepsilon \to 0} T_\varepsilon(u),$$

uniformly for $u \in D$. Finally, take $T_k := T_\varepsilon$ with $\varepsilon = 1/k$. ■

Here is a notion whose definition involves completely continuous operators.

Definition 2.2 Let $(X, |.|_X)$ and $(Y, |.|_Y)$ be Banach spaces such that $Y \subset X$. We say that the embedding $Y \subset X$ is *continuous* (respectively, *completely continuous*) if the injection map $j : Y \to X$, $j(u) = u$ $(u \in Y)$ is continuous, respectively, completely continuous.

We note that in the literature, the term '*compact embedding*' is frequently used instead of 'completely continuous embedding'.

It is clear that the embedding $Y \subset X$ is continuous if and only if there is a constant $c > 0$ such that

$$|u|_X \leq c\,|u|_Y\,, \quad u \in Y.$$

For example, according to Remark 1.1 the embedding $C^k\left(\overline{\Omega}; \mathbf{R}^n\right) \subset C^{k-1}\left(\overline{\Omega}; \mathbf{R}^n\right)$ is completely continuous. Here $\Omega \subset \mathbf{R}^N$ is bounded open and $k \in \mathbf{N} \setminus \{0\}$. Also, if $1 \leq p < q \leq \infty$ and Ω is bounded open then the embedding $L^q(\Omega; \mathbf{R}^n) \subset L^p(\Omega; \mathbf{R}^n)$ is continuous since by Hölder's inequality

$$|u|_p \leq \mu(\Omega)^{\frac{q-p}{pq}}\,|u|_q\,, \quad u \in L^q(\Omega; \mathbf{R}^n)\,.$$

Here $(q-p)/(pq) = 1/p$ if $q = \infty$.

Note that if the embedding $Y \subset X$ is continuous (respectively, completely continuous), then the embedding $X^* \subset Y^*$ is continuous (respectively, completely continuous).

2.2 Brouwer's Fixed Point Theorem

The basic topological tool in Part I will be Schauder's fixed point theorem. In this section we prove the finite-dimensional version of Schauder's theorem, namely Brouwer's fixed point theorem [8].

Theorem 2.3 (Brouwer) *Let $D \subset \mathbf{R}^N$ be a nonempty convex compact set and let $T : D \to D$ be a continuous mapping. Then there exists at least one $u \in D$ with $T(u) = u$.*

For the proof we need the following lemmas:

Lemma 2.1 *Let $F : \mathbf{R}^{n+1} \to \mathbf{R}^n$ be a C^∞ function. Consider the determinant*

$$A_i = \det\left[D_0F\,,\, ...,\, D_{i-1}F\,,\, D_{i+1}F\,,\, ...,\, D_nF\right]$$

whose columns are

$$D_iF = \left(\partial F_1/\partial x_{i+1},\, ...,\, \partial F_n/\partial x_{i+1}\right).$$

Then

$$\sum_{i=0}^{n} (-1)^i\, D_iA_i = 0, \tag{2.2}$$

where $D_iA_i = \partial A_i/\partial x_{i+1}$.

Proof. Let $i, j \in \{0, 1, ..., n\}$, $i \neq j$. For $i < j$ we define

$$C_{ij} = \det [D_{ij}F, D_0F, ..., D_{i-1}F, D_{i+1}F, ..., D_{j-1}F, D_{j+1}F, ..., D_nF]$$

where

$$D_{ij}F = \left(\partial^2 F_1/\partial x_{i+1}\partial x_{j+1}, ..., \partial^2 F_n/\partial x_{i+1}\partial x_{j+1}\right).$$

For $i > j$ we let $C_{ij} = C_{ji}$. We can easily check that

$$D_i A_i = \sum_{j<i} (-1)^j C_{ij} + \sum_{j>i} (-1)^{j-1} C_{ij} = \sum_{j=0}^{n} (-1)^j \sigma(i,j) C_{ij}$$

where

$$\sigma(i,j) = \begin{cases} 1 & \text{if } j < i, \\ 0 & \text{if } j = i, \\ -1 & \text{if } j > i. \end{cases}$$

Hence

$$\sum_{i=0}^{n} (-1)^i D_i A_i = \sum_{i,j=0}^{n} (-1)^{i+j} \sigma(i,j) C_{ij}.$$

Since $\sigma(i,j) = -\sigma(j,i)$ and $C_{ij} = C_{ji}$, we have

$$\begin{aligned} \sum_{i,j=0}^{n} (-1)^{i+j} \sigma(i,j) C_{ij} &= \sum_{i,j=0}^{n} (-1)^{i+j} \sigma(j,i) C_{ji} \\ &= -\sum_{i,j=0}^{n} (-1)^{i+j} \sigma(i,j) C_{ij}, \end{aligned}$$

which proves (2.2). ■

The next lemma is Brouwer's theorem for mappings of class C^∞.

Lemma 2.2 *Let $T : B \to B$, where $B = \overline{B}_1(0; \mathbf{R}^n)$, be a C^∞ function. Then T has at least one fixed point.*

Proof. Assume that T has no fixed points. Then for any $u \in B$ the line generating by u and $T(u)$:

$$v = u + \alpha(u - T(u)), \quad \alpha \in \mathbf{R},$$

cuts the sphere ∂B in two points. Let $\alpha(u)$ be the nonnegative value of α which corresponds to one of these two points. It is clear that $\alpha(u)$ is the biggest root of the equation

$$|u + \alpha(u - T(u))| = 1,$$

which can be rewritten in the equivalent form

$$\alpha^2 |u - T(u)|^2 + 2\alpha (u, u - T(u)) + |u|^2 = 1.$$

Hence

$$\begin{aligned}\alpha(u) &= |u - T(u)|^{-2} \{(u, T(u) - u) \\ &\quad + \left((u, u - T(u))^2 + (1 - |u|^2) |u - T(u)|^2\right)^{1/2}\}.\end{aligned}$$

This defines a C^∞ function from B to $\mathbf{R}$. In addition,

$$\alpha(u) = 0 \quad \text{for} \quad |u| = 1.$$

Let us now define the mapping $F : [0,1] \times B \to B$ by

$$F(t, u) = u + t\,\alpha(u)(u - T(u)).$$

We have that F is of class C^∞ and satisfies

$$\begin{cases} D_t F(t, u) = 0 \quad \text{for} \quad |u| = 1, \\ F(0, u) = u, \\ |F(1, u)| = 1. \end{cases} \tag{2.3}$$

Here $D_t F(t, u) = \partial F(t, u) / \partial t$. Let

$$A_0(t, u) = \det[D_1 F(t, u), ..., D_n F(t, u)]$$

and

$$I(t) = \int_B A_0(t, u)\, du.$$

We have $A_0(0, u) = 1$ and $I(0) = \mu(B) > 0$. Also $A_0(1, u) \equiv 0$. Indeed, the vectors $D_1 F(1, u), ..., D_n F(1, u)$ belong to the tangent hyperplane to ∂B at $F(1, u)$. Thus they are linearly dependent and so $A_0(1, u) \equiv 0$, as we claimed. Consequently $I(1) = 0$. We obtain a contradiction once we prove that $I(t)$ is constant; more exactly, that $I'(t) \equiv 0$. To this end we differentiate under the integral sign and we use (2.2). We then obtain

$$I'(t) = \int_B D_t A_0(t, u)\, du = \sum_{i=1}^{n} (-1)^{i+1} \int_B D_i A_i(t, u)\, du,$$

where

$$A_i(t, u) = \det[D_t F, D_1 F, ..., D_{i-1} F, D_{i+1} F, ..., D_n F].$$

Let $B_i = B \cap \{u \in \mathbf{R}^n : u_i = 0\}$. For each $u \in B_i$ we define

$$\varphi_i^+(u) = u + (0, ..., \gamma_i(u), ..., 0)$$

and

$$\varphi_i^-(u) = u + (0, ..., -\gamma_i(u), ..., 0)$$

where $\gamma_i(u) = (1 - \sum_{j \neq i} u_j^2)^{1/2}$. Then

$$\begin{aligned}\int_B D_i A_i(t,u)\,du &= \int_{B_i} \int_{-\gamma_i(u)}^{\gamma_i(u)} D_i A_i(t,u)\,du \\ &= \int_{B_i} \left[A_i(t, \varphi_i^+(u)) - A_i(t, \varphi_i^-(u))\right] \\ &\quad \times du_1 \ldots du_{i-1} du_{i+1} \ldots du_n \\ &= 0\end{aligned}$$

for $i = 1, 2, ..., n$. Indeed, since $|\varphi_i^{\pm}(u)| = 1$, according to (2.3), one has $D_t F(t, \varphi_i^{\pm}(u)) = 0$. Hence the first column of the determinant $A_i(t, \varphi_i^{\pm}(u))$ is null, and so this determinant equals zero. Therefore $I'(t) \equiv 0$. ∎

Proof of Theorem 2.3. *Step* 1. Assume that $D = B = \overline{B}_1(0; \mathbf{R}^n)$. Let

$$T(u) = (t_1(u), t_2(u), ..., t_n(u))$$

where $t_i : B \to [-1, 1]$, $i = 1, 2, ..., n$, are continuous functions. According to Weierstrass' approximation theorem, each function t_i is the uniform limit of a sequence $(t_i^k)_{k \geq 1}$ of functions $t_i^k : B \to [-1, 1]$ of class C^∞ satisfying

$$\left|t_i^k(u)\right| \leq |t_i(u)|$$

for all $u \in B$, $i = 1, 2, ..., n$ and $k = 1, 2, ...$ Define $T_k : B \to B$ by

$$T_k(u) = \left(t_1^k(u), t_2^k(u), ..., t_n^k(u)\right).$$

Then

$$T(u) = \lim_{k \to \infty} T_k(u)$$

uniformly on B. By Lemma 2.2, for each k there exists a $u_k \in B$ with

$$T_k(u_k) = u_k.$$

Since B is compact, (u_k) has a subsequence $\left(u_{k_j}\right)_{j\geq 1}$ convergent to some element $u_0 \in B$. Then

$$\begin{aligned} |T(u_0) - u_0| &\leq \left|T(u_0) - T\left(u_{k_j}\right)\right| + \left|T\left(u_{k_j}\right) - T_{k_j}\left(u_{k_j}\right)\right| \\ + \left|u_{k_j} - u_0\right| &\rightarrow 0 \quad \text{as } j \rightarrow \infty. \end{aligned}$$

Therefore $T(u_0) = u_0$.

Step 2. The case $D = \overline{B}_r(0; \mathbf{R}^n)$ can be reduced to the previous one by the change of variables $v = r^{-1}u$ and by considering the operator $T' : B \rightarrow B$,

$$T'(v) = r^{-1}T(rv) \quad (v \in B).$$

Step 3. We now consider the general case. We first show that there is a continuous extension $\widetilde{T}$ of T to the whole space $\mathbf{R}^n$ with $\widetilde{T}(\mathbf{R}^n) \subset D$. Indeed, D being compact in $\mathbf{R}^n$, we may choose a countable, dense subset $\{u_j : j \in \mathbf{N} \setminus \{0\}\}$ of D. For any $u \notin D$ and any j we let

$$\psi_j(u) = \max\left\{2 - \frac{|u - u_j|}{d(u, D)}, 0\right\}.$$

Here

$$d(u, D) = \inf\{|u - v| : v \in D\}.$$

Then the mapping $\widetilde{T} : \mathbf{R}^n \rightarrow D$ given by

$$\widetilde{T}(u) = \begin{cases} T(u) & \text{if } u \in D, \\ \left(\sum\limits_{j\geq 1} 2^{-j}\psi_j(u)\right)^{-1} \sum\limits_{j\geq 1} 2^{-j}\psi_j(u)T(u_j) & \text{if } u \notin D \end{cases}$$

is the desired continuous extension of T. Now choose any $r > 0$ large enough so that $D \subset \overline{B}_r(0; \mathbf{R}^n)$ and define

$$S : \overline{B}_r(0; \mathbf{R}^n) \rightarrow D \subset \overline{B}_r(0; \mathbf{R}^n), \quad S(u) = \widetilde{T}(u).$$

We are with S in the case considered at Step 2. Hence there exists a $u \in \overline{B}_r(0; \mathbf{R}^n)$ with $S(u) = u$, that is $\widetilde{T}(u) = u$. Since $\widetilde{T}(\mathbf{R}^n) \subset D$, we have $u \in D$ and so $T(u) = u$. ■

For another proof of Brouwer's theorem see Kantorovitch–Akilov [29].

2.3 Schauder's Fixed Point Theorem

The main result in this chapter is the famous fixed point theorem of Schauder [49].

Theorem 2.4 (Schauder) *Let X be a Banach space, $K \subset X$ a nonempty convex compact set and let $T : K \to K$ be a continuous operator. Then T has at least one fixed point.*

Proof. Obviously T is completely continuous. Consequently by Theorem 2.2 there exists a sequence of continuous operators $T_j : K \to K$ of finite rank such that

$$T(u) = \lim_{j \to \infty} T_j(u)$$

uniformly on K. Let $n = n(j)$ be the dimension of the subspace X_n generated by $T_j(K)$. We have

$$T_j : K \cap X_n \to K \cap X_n.$$

Consequently by Brouwer's theorem there exists $u_j \in K \cap X_n$ with

$$T_j(u_j) = u_j.$$

Since K is compact there exists a subsequence of (u_j) convergent to some element $u \in K$. As in Step 1 in the proof of Brouwer's theorem we can conclude that $T(u) = u$. ■

The following variant of Schauder's theorem is most useful in applications.

Theorem 2.5 (Schauder) *Let X be a Banach space, $D \subset X$ a nonempty convex bounded closed set and let $T : D \to D$ be a completely continuous operator. Then T has at least one fixed point.*

We can derive Theorem 2.5 from Theorem 2.4 via the following result.

Lemma 2.3 (Mazur) *The convex hull of any relatively compact subset of a Banach space is relatively compact.*

Proof. Let Y be a relatively compact subset of a Banach space X. Then, given $\varepsilon > 0$ we find a finite number of elements of X, say $u_1, u_2, ..., u_m$ such that

$$Y \subset \bigcup_{i=1}^{m} B_\varepsilon(u_i). \tag{2.4}$$

Let

$$\mathcal{R} = \operatorname{conv}\{u_1, u_2, ..., u_m\}.$$

Our goal is to prove that $\mathcal{R}$ is a relatively compact ε-net for $\operatorname{conv}(Y)$. Once this is proved, we may say that $\operatorname{conv}(Y)$ is a relatively compact set in base of Hausdorff's theorem. To this end consider an arbitrary element $u \in \operatorname{conv}(Y)$. We have

$$u = \sum_{j=1}^{n} \lambda_j v_j, \quad \lambda_j > 0, \quad \sum_{j=1}^{n} \lambda_j = 1, \quad v_j \in Y.$$

According to (2.4), for each v_j there is an $i_j \in \{1, 2, ..., m\}$ with $v_j \in B_\varepsilon(u_{i_j})$. Then

$$\begin{aligned} \left|u - \sum_{j=1}^{n} \lambda_j u_{i_j}\right| &= \left|\sum_{j=1}^{n} \lambda_j \left(v_j - u_{i_j}\right)\right| \\ &\leq \sum_{j=1}^{n} \lambda_j \left|v_j - u_{i_j}\right| \\ &< \varepsilon \end{aligned}$$

and $\sum_{j=1}^{n} \lambda_j u_{i_j} \in \mathcal{R}$. This shows that $\mathcal{R}$ is an ε-net for $\operatorname{conv}(Y)$. Finally, the compactness of $\mathcal{R}$ follows from the representation:

$$\mathcal{R} = \left\{ \sum_{j=1}^{m} \lambda_j u_j : 0 \leq \lambda_j \leq 1, \ \sum_{j=1}^{m} \lambda_j = 1 \right\}.$$

Thus the proof is complete. ■

Proof of Theorem 2.5. Since T is completely continuous and D is bounded the set $T(D)$ is relatively compact. Mazur's lemma then implies that the set $K = \overline{\operatorname{conv}}\,(T(D))$ is compact (and obviously convex). From $T(D) \subset D$ and D closed convex it follows that $K \subset D$. Now we apply Theorem 2.4 to the operator $T : K \to K$. ■

Chapter 3

Continuous Solutions of Integral Equations via Schauder's Theorem

This chapter presents three examples of nonlinear integral operators which are completely continuous on some spaces of continuous functions: the Fredholm integral operator, the Volterra integral operator, and a particular integral operator with delay. Simultaneously, by means of Schauder's fixed point theorem we prove existence theorems for continuous solutions of the integral equations associated to these operators.

3.1 The Fredholm Integral Operator

Let $\Omega \subset \mathbf{R}^N$ be a bounded open set.

Theorem 3.1 *Let $h : \overline{\Omega}^2 \times \mathbf{R}^n \to \mathbf{R}^n$ be a continuous mapping. Then the Fredholm operator associated to h, $T : C\left(\overline{\Omega}; \mathbf{R}^n\right) \to C\left(\overline{\Omega}; \mathbf{R}^n\right)$ given by*

$$T(u)(x) = \int_{\Omega} h(x, y, u(y))\, dy \quad \left(x \in \overline{\Omega}\right) \tag{3.1}$$

is completely continuous.

Proof. We first prove that T is continuous. Let $u_0 \in C\left(\overline{\Omega}; \mathbf{R}^n\right)$ and choose any number $R > |u_0|_\infty$. Let $\varepsilon > 0$. Since h is uniformly continuous on the compact set $\overline{\Omega}^2 \times \overline{B}_R(0; \mathbf{R}^n)$, there exists a $\delta_\varepsilon > 0$ such that for every $u \in C(\overline{\Omega}; \mathbf{R}^n)$ satisfying $|u - u_0|_\infty \le \delta_\varepsilon$ one has $u(y) \in \overline{B}_R(0; \mathbf{R}^n)$ and

$$|h(x, y, u(y)) - h(x, y, u_0(y))| \le \varepsilon$$

for all $x, y \in \overline{\Omega}$. Then

$$\begin{aligned} |T(u)(x) - T(u_0)(x)| &\le \int_\Omega |h(x, y, u(y)) - h(x, y, u_0(y))| \, dy \\ &\le \varepsilon \, \mu(\Omega) \end{aligned}$$

for every $x \in \overline{\Omega}$. Hence

$$|T(u) - T(u_0)|_\infty \le \varepsilon \, \mu(\Omega)$$

whenever $|u - u_0|_\infty \le \delta_\varepsilon$. Therefore T is continuous at u_0.

Next, given a bounded subset Y of $C(\overline{\Omega}; \mathbf{R}^n)$, we shall prove that $T(Y)$ is relatively compact in $C(\overline{\Omega}; \mathbf{R}^n)$. According to the Ascoli–Arzèla theorem, we have to show that $T(Y)$ is bounded and equicontinuous.

Indeed, since Y is bounded there exists a constant $c > 0$ such that

$$|u|_\infty \le c \quad \text{for all } u \in Y.$$

It follows that for any $u \in Y$ we have

$$|T(u)|_\infty \le M \, \mu(\Omega),$$

where

$$M = \max_{\overline{\Omega}^2 \times \overline{B}_c(0; \mathbf{R}^n)} |h(x, y, z)|.$$

Hence the set $T(Y)$ is bounded in $C(\overline{\Omega}; \mathbf{R}^n)$.

On the other hand, using the uniform continuity of h on the compact $\overline{\Omega}^2 \times \overline{B}_c(0; \mathbf{R}^n)$, for each $\varepsilon > 0$ there exists a $\delta_\varepsilon > 0$ such that

$$|h(x, y, u(y)) - h(x', y, u(y))| \le \varepsilon$$

for all $x, x', y \in \overline{\Omega}$ with $|x - x'| \le \delta_\varepsilon$ and $u \in Y$. This immediately yields

$$|T(u)(x) - T(u)(x')| \le \varepsilon \, \mu(\Omega),$$

for all $x, x' \in \overline{\Omega}$ satisfying $|x - x'| \le \delta_\varepsilon$ and $u \in Y$. Thus $T(Y)$ is equicontinuous. ∎

The next result is a local version of Theorem 3.1.

Theorem 3.2 *Let* $R > 0$ *and* $h : \overline{\Omega}^2 \times \overline{B}_R(0; \mathbf{R}^n) \to \mathbf{R}^n$ *be a continuous mapping. Then the operator* $T : \overline{B}_R\left(0; C\left(\overline{\Omega}; \mathbf{R}^n\right)\right) \to C\left(\overline{\Omega}; \mathbf{R}^n\right)$ *given by* (3.1) *is completely continuous.*

Proof. Essentially the same reasoning as in the proof of Theorem 3.1 establishes the result. ■

Using Theorem 3.2 and Schauder's fixed point theorem we shall prove the existence of continuous solutions in a ball of $C\left(\overline{\Omega}; \mathbf{R}^n\right)$, to the *Fredholm integral equation* in $\mathbf{R}^n$

$$u(x) = \int_\Omega h(x, y, u(y))\, dy, \quad x \in \overline{\Omega}. \tag{3.2}$$

Theorem 3.3 *Let* $R > 0$ *and* $h : \overline{\Omega}^2 \times \overline{B}_R(0; \mathbf{R}^n) \to \mathbf{R}^n$ *be a continuous mapping. Assume*

$$M\, \mu(\Omega) \leq R, \tag{3.3}$$

where

$$M = \max_{\overline{\Omega}^2 \times \overline{B}_R(0; \mathbf{R}^n)} |h(x, y, z)|.$$

Then (3.2) *has at least one solution* $u \in C\left(\overline{\Omega}; \mathbf{R}^n\right)$ *with* $|u|_\infty \leq R$.

Proof. According to Theorem 3.2, the operator $T : \overline{B}_R\left(0; C\left(\overline{\Omega}; \mathbf{R}^n\right)\right) \to C\left(\overline{\Omega}; \mathbf{R}^n\right)$ given by (3.1) is completely continuous. On the other hand, (3.3) guarantees that

$$T\left(\overline{B}_R\left(0; C\left(\overline{\Omega}; \mathbf{R}^n\right)\right)\right) \subset \overline{B}_R\left(0; C\left(\overline{\Omega}; \mathbf{R}^n\right)\right).$$

Thus the conclusion follows from Theorem 2.5. ■

3.2 The Volterra Integral Operator

Essentially the same reasoning as in the proof of Theorem 3.1 establishes the following two results:

Theorem 3.4 *Let* $h : [a, b]^2 \times \mathbf{R}^n \to \mathbf{R}^n$ *be continuous. Then the Volterra operator associated to* h, $T : C([a, b]; \mathbf{R}^n) \to C([a, b]; \mathbf{R}^n)$ *given by*

$$T(u)(t) = \int_a^t h(t, s, u(s))\, ds \quad (t \in [a, b]) \tag{3.4}$$

is completely continuous.

Theorem 3.5 *Let $R > 0$ and let $h : [a,b]^2 \times \overline{B}_R(0;\mathbf{R}^n) \to \mathbf{R}^n$ be continuous. Then the operator $T : \overline{B}_R(0; C([a,b];\mathbf{R}^n)) \to C([a,b];\mathbf{R}^n)$ given by (3.4) is completely continuous.*

As an application we present an existence theorem for the *Volterra integral equation* in $\mathbf{R}^n$

$$u(t) = \int_a^t h(t,s,u(s))\,ds + v(t), \quad t \in [a,b]. \tag{3.5}$$

Theorem 3.6 *Let $h : [a,b]^2 \times \mathbf{R}^n \to \mathbf{R}^n$ be continuous and let $v \in C([a,b];\mathbf{R}^n)$. Assume that there exists constants $\alpha,\ \beta \in \mathbf{R}_+$ such that*

$$|h(t,s,z)| \le \alpha |z| + \beta \tag{3.6}$$

for all $t,\ s \in [a,b]$, $z \in \mathbf{R}^n$. Then (3.5) has at least one solution $u \in C([a,b];\mathbf{R}^n)$.

Proof. Let $T : C([a,b];\mathbf{R}^n) \to C([a,b];\mathbf{R}^n)$ be given by

$$T(u)(t) = \int_a^t h(t,s,u(s))\,ds + v(t).$$

According to Theorem 3.4, T is completely continuous. We now show that T is a self-mapping of a closed ball of the space $C([a,b];\mathbf{R}^n)$ endowed with a suitable norm, equivalent to the sup-norm $|.|_\infty$. Indeed, for any given number $\theta > 0$ we have

$$\begin{aligned}
|T(u)(t)| &= \left|\int_a^t h(t,s,u(s))\,ds + v(t)\right| \\
&\le \int_a^t |h(t,s,u(s))|\,ds + |v|_\infty \\
&\le \alpha \int_a^t |u(s)|\,ds + \beta\,(b-a) + |v|_\infty \\
&= \alpha \int_a^t |u(s)|\, e^{-\theta(s-a)} e^{\theta(s-a)} ds + \beta(b-a) + |v|_\infty \\
&\le \alpha \left|u(.)\,e^{-\theta(.-a)}\right|_\infty \int_a^t e^{\theta(s-a)} ds + \beta(b-a) + |v|_\infty \\
&= \alpha\theta^{-1} \left|u(.)\,e^{-\theta(.-a)}\right|_\infty \left(e^{\theta(t-a)} - 1\right) + \beta(b-a) + |v|_\infty \\
&\le \alpha\theta^{-1} \left|u(.)\,e^{-\theta(.-a)}\right|_\infty e^{\theta(t-a)} + \beta(b-a) + |v|_\infty
\end{aligned}$$

Since $e^{-\theta(t-a)} \le 1$ on $[a,b]$, we deduce

$$|T(u)(t)|\, e^{-\theta(t-a)} \le \alpha\,\theta^{-1} \left|u(.)\, e^{-\theta(.-a)}\right|_\infty + \beta(b-a) + |v|_\infty$$

and so

$$\left|T(u)(.)\, e^{-\theta(.-a)}\right|_\infty \le \alpha\,\theta^{-1} \left|u(.)\, e^{-\theta(.-a)}\right|_\infty + \beta(b-a) + |v|_\infty . \tag{3.7}$$

Now fix any $\theta > \alpha$. Then we can find $R > 0$ such that

$$\alpha\,\theta^{-1} R + \beta(b-a) + |v|_\infty \le R. \tag{3.8}$$

Consider a new norm on $C([a,b];\mathbf{R}^n)$, namely

$$\|u\| = \left|u(.)\, e^{-\theta(.-a)}\right|_\infty .$$

It is clear that the norm $|.|_\infty$ and $\|.\|$ are equivalent (thus T is also completely continuous with respect to $\|.\|$). On the other hand, (3.7) and (3.8) show that T maps the closed ball of center 0 and radius R of the space $(C([a,b];\mathbf{R}^n), \|.\|)$, into itself. Now the conclusion follows from Theorem 2.5. ■

The reader could try to give variants of Theorem 3.6 to the case where condition (3.6) is replaced by an inequality of the form

$$|h(t,s,z)| \le \psi(|z|)$$

with other types of functions $\psi : \mathbf{R}_+ \to \mathbf{R}_+$.

3.3 An Integral Operator with Delay

The following *delay integral equation*

$$u(t) = \int_{t-\tau}^{t} f(s, u(s))\, ds$$

can be interpreted as a model for the spread of certain infectious diseases with a contact rate that varies seasonally. In this equation $u(t)$ is the proportion of infectives in a population at time t, τ is the length of time an individual remains infectious, and $f(t, u(t))$ is the proportion of new infectives per unit time.

In this section we study the existence of continuous solutions on a given interval of time $[0, t_1]$, for the initial value problem

$$\begin{cases} u(t) = \int_{t-\tau}^{t} f(s, u(s))\, ds, & 0 \leq t \leq t_1, \\ u(t) = \varphi(t), & -\tau \leq t \leq 0. \end{cases} \tag{3.9}$$

We assume

$$f \in C([-\tau, t_1] \times \mathbf{R}^n; \mathbf{R}^n), \quad \varphi \in C([-\tau, 0]; \mathbf{R}^n)$$

and the following sewing condition holds

$$\varphi(0) = \int_{-\tau}^{0} f(s, \varphi(s))\, ds. \tag{3.10}$$

By a solution of (3.9) we mean a function $u \in C([-\tau, t_1]; \mathbf{R}^n)$ with $u(t) = \varphi(t)$ for all $t \in [-\tau, 0]$.

The initial value problem (3.9) was studied for the first time in Precup [44].

Theorem 3.7 *Assume* $f \in C([-\tau, t_1] \times \mathbf{R}^n; \mathbf{R}^n)$, $\varphi \in C([-\tau, 0]; \mathbf{R}^n)$ *and that* (3.10) *holds. Then the delay integral operator* $T : D(T) \to C([0, t_1]; \mathbf{R}^n)$ *given by*

$$T(u)(t) = \int_{t-\tau}^{t} f(s, \widetilde{u}(s))\, ds \quad (t \in [0, t_1]),$$

where

$$D(T) = \{u \in C([0, t_1]; \mathbf{R}^n) : u(0) = \varphi(0)\}$$

and

$$\widetilde{u}(t) = \begin{cases} \varphi(t) & \text{for } t \in [-\tau, 0], \\ u(t) & \text{for } t \in [0, t_1], \end{cases}$$

is completely continuous.

Proof. Use the Ascoli–Arzèla theorem and follow the same steps as in the proof of Theorem 3.1. We omit the details. ■

Theorem 3.8 *Assume* $f \in C([-\tau, t_1] \times \mathbf{R}^n; \mathbf{R}^n)$, $\varphi \in C([-\tau, 0]; \mathbf{R}^n)$ *and that* (3.10) *holds. In addition assume that there exist* α, $\beta \in \mathbf{R}_+$ *such that*

$$|f(t, z)| \leq \alpha\, |z| + \beta \tag{3.11}$$

for all $z \in \mathbf{R}^n$ *and* $t \in [0, t_1]$. *Then* (3.9) *has a solution* $u \in C([-\tau, t_1]; \mathbf{R}^n)$.

Proof. The proof closely resembles the proof of Theorem 3.6. Let

$$\gamma = \max_{t \in [-\tau, 0]} |f(t, \varphi(t))|.$$

We have

$$\begin{aligned} |T(u)(t)| &\leq \tau\gamma + \int_0^t |f(s, u(s))| \, ds \\ &\leq \tau\gamma + \alpha \int_0^t |u(s)| \, ds + \beta t_1 \\ &= \tau\gamma + \beta t_1 + \alpha \int_0^t |u(s)| e^{-\theta s} e^{\theta s} ds \\ &\leq \tau\gamma + \beta t_1 + \alpha \left| u(s) e^{-\theta s} \right|_\infty \int_0^t e^{\theta s} ds \\ &\leq \tau\gamma + \beta t_1 + \alpha \left| u(s) e^{-\theta s} \right|_\infty \theta^{-1} e^{\theta t}. \end{aligned}$$

It follows that

$$\left| T(u)(t) e^{-\theta t} \right|_\infty \leq \tau\gamma + \beta t_1 + \alpha \theta^{-1} \left| u(s) e^{-\theta s} \right|_\infty.$$

Now choose $\theta > \alpha$ and a number $R > 0$ with

$$\tau\gamma + \beta t_1 + \alpha \theta^{-1} R \leq R.$$

Then $T(B) \subset B$, where

$$B = \left\{ u \in D(T) : \left| u(t) e^{-\theta t} \right| \leq R \text{ for all } t \in [0, t_1] \right\}.$$

It is clear that B is a nonempty convex bounded closed subset of $C([0, t_1]; \mathbf{R}^n)$. The conclusion is now immediate from Theorem 2.5. ■

As in the previous section, the reader could try to obtain existence results for (3.9) assuming instead of (3.11) a condition of the form

$$|f(t, z)| \leq \psi(|z|)$$

with different types of function $\psi : \mathbf{R}_+ \to \mathbf{R}_+$.

Chapter 4

The Leray–Schauder Principle and Applications

In applications one of the drawbacks of Schauder's fixed point theorem is the invariance condition $T(D) \subset D$ which has to be guaranteed for a bounded closed convex subset D of a Banach space. The Leray–Schauder principle [32] makes it possible to avoid such a condition and requires instead that a 'boundary condition' is satisfied. In this chapter we shall prove the Leray–Schauder principle and we shall apply it in order to obtain existence results for continuous solutions of integral equations. In particular, we give results on the existence of continuous solutions of initial value and two-point boundary value problems for nonlinear ordinary differential equations in $\mathbf{R}^n$. The results will be better than those established by means of Schauder's theorem.

4.1 The Leray–Schauder Principle

Theorem 4.1 (Leray–Schauder) *Let X be a Banach space, $K \subset X$ a closed convex subset, $U \subset K$ a bounded set, open in K and $u_0 \in U$ a fixed element. Assume that the operator $T : \overline{U} \to K$ is completely continuous and satisfies the boundary condition*

$$u \neq (1-\lambda)\,u_0 + \lambda T(u) \quad \text{for all } u \in \partial U,\ \lambda \in (0,1)\,. \tag{4.1}$$

Then T has at least one fixed point in $\overline{U}$.

Proof. For the proof we use Granas' fixed point approach (see Dugundji–Granas [19], Granas [23], O'Regan–Precup [37] and Zeidler [53]). Notice the property of U of being open as well as the boundary ∂U are understood with respect to the topology of K. We may assume that (4.1) holds on ∂U for all $\lambda \in [0,1]$. Indeed, this is obvious for $\lambda = 0$ since $u_0 \in U$, whilst if (4.1) does not hold for $\lambda = 1$, then the theorem is proved.

Let

$$\mathcal{S} = \left\{ u \in \overline{U} : u = (1-\lambda)\, u_0 + \lambda T(u) \text{ for some } \lambda \in [0,1] \right\}.$$

Obviously $\mathcal{S}$ is nonempty (since $u_0 \in \mathcal{S}$), closed, and $\mathcal{S} \cap \partial U = \emptyset$. By Urysohn's lemma (see Dugundji–Granas [19]), there exists a function $\varphi \in C\left(\overline{U}; [0,1]\right)$ such that

$$\varphi(u) = \begin{cases} 0 & \text{for } u \in \partial U, \\ 1 & \text{for } u \in \mathcal{S}. \end{cases}$$

We now define the operator $\widetilde{T} : K \to K$ by

$$\widetilde{T}(u) = \begin{cases} (1-\varphi(u))\, u_0 + \varphi(u)\, T(u) & \text{for } u \in U, \\ u_0 & \text{for } u \in K \setminus U. \end{cases}$$

It is clear that $\widetilde{T}$ is continuous and

$$\widetilde{T}(K) \subset \operatorname{conv}\left(\{u_0\} \cup T\left(\overline{U}\right)\right).$$

Since T is completely continuous, $T\left(\overline{U}\right)$ is relatively compact. Hence by Mazur's lemma the following subset of K,

$$D = \overline{\operatorname{conv}}\left(\{u_0\} \cup T\left(\overline{U}\right)\right),$$

is convex and compact. In addition

$$\widetilde{T}(D) \subset D.$$

Hence Theorems 2.4 applies and guarantees the existence of a $u \in D$ with $\widetilde{T}(u) = u$. By the definition of $\widetilde{T}$, u must lie in U. Then

$$u = (1-\varphi(u))\, u_0 + \varphi(u)\, T(u).$$

This shows that $u \in \mathcal{S}$ and so $\varphi(u) = 1$. As a result, $u = T(u)$. ■

Notice that the essential idea of the Leray–Schauder principle consists in joining the operator T to the constant operator u_0 by means of the homotopy $H : \overline{U} \times [0,1] \to K$,

$$H(u,\lambda) = (1-\lambda)\, u_0 + \lambda T(u)$$

in a such way that the unique fixed point of $H(.,0)$, namely u_0, can be 'continued' in a fixed point of $H(.,\lambda)$ for each $\lambda \in [0,1]$, and, in particular, in a fixed point of $H(.,1) = T$. This continuation process is possible if all operators $H(.,\lambda)$ for $\lambda \in [0,1]$ are fixed point free on the boundary of U.

In applications the Leray–Schauder principle is usually used together with the so called '*a priori* bounds technique':

Suppose we wish to solve the operator equation

$$u = T(u), \quad u \in K, \tag{4.2}$$

where K is a closed, convex subset of a Banach space $(X, |.|_X)$, and $T : K \to K$ is completely continuous. Then we look at the set of all solutions to the one-parameter family of equations

$$u = (1-\lambda)\,u_0 + \lambda T(u), \quad u \in K, \tag{4.3}$$

when $\lambda \in (0,1)$. Here $u_0 \in K$ is fixed (in most cases $u_0 = 0$). If this set is bounded, i.e., there exists $R > 0$ such that

$$|u - u_0|_X < R$$

(strictly!) whenever u solves (4.3) for some $\lambda \in (0,1)$, then we let U be the intersection of K with the open ball $B_R(u_0; X)$. Thus Theorem 4.1 applies and guarantees the existence of a solution to (4.2).

For other results of Leray–Schauder type we refer the reader to the monograph O'Regan–Precup [37].

4.2 Existence Results for Fredholm Integral Equations

In this section we present general existence theorems for the Fredholm integral equation in $\mathbf{R}^n$

$$u(x) = \int_\Omega h(x, y, u(y))\,dy, \quad x \in \overline{\Omega}. \tag{4.4}$$

Here $\Omega \subset \mathbf{R}^N$ is a bounded open set. We seek continuous solutions with values in a given ball

$$B = \{z \in \mathbf{R}^n : |z| \le R\},$$

i.e., $u \in C(\overline{\Omega}; B)$.

Theorem 4.1 yields the following existence principle which can be summarized as follows: 'boundedness yields existence'.

Theorem 4.2 *Let* $h : \overline{\Omega}^2 \times B \to \mathbf{R}^n$ *be continuous. Assume that*

$$|u|_\infty < R \tag{4.5}$$

for any solution $u \in C\left(\overline{\Omega}; B\right)$ *to*

$$u(x) = \lambda \int_\Omega h(x, y, u(y))\, dy, \quad x \in \overline{\Omega}, \tag{4.6}$$

for each $\lambda \in (0,1)$. *Then* (4.4) *has a solution in* $C\left(\overline{\Omega}; B\right)$.

Proof. Let $K = X = C\left(\overline{\Omega}; \mathbf{R}^n\right)$ with norm $|.|_\infty$,

$$U = \{u \in C\left(\overline{\Omega}; \mathbf{R}^n\right) : |u|_\infty < R\},$$

u_0 be the null function and $T : \overline{U} \to C\left(\overline{\Omega}; \mathbf{R}^n\right)$ be given by

$$T(u)(x) = \int_\Omega h(x, y, u(y))\, dy \quad \left(x \in \overline{\Omega}\right).$$

The result follows from Theorem 4.1. ■

Notice Theorem 3.3 is a corollary of Theorem 4.2. Indeed, if $u \in C\left(\overline{\Omega}; B\right)$ is any solution of (4.6) for some $\lambda \in (0,1)$, then for any $x \in \overline{\Omega}$, using (3.3) we obtain

$$\begin{aligned} |u(x)| &= \lambda \left| \int_\Omega h(x, y, u(y))\, dy \right| \\ &\leq \lambda \int_\Omega |h(x, y, u(y))|\, dy \\ &\leq \lambda M \mu(\Omega) \\ &\leq \lambda R < R. \end{aligned}$$

Hence u satisfies (4.5).

An immediate consequence of Theorem 4.2 is the following existence result for the *Hammerstein integral equation* in $\mathbf{R}^n$

$$u(x) = \int_\Omega \kappa(x, y) f(y, u(y))\, dy, \quad x \in \overline{\Omega}. \tag{4.7}$$

Corollary 4.1 *Let* $\kappa : \overline{\Omega}^2 \to \mathbf{R}$ *and* $f : \overline{\Omega} \times B \to \mathbf{R}^n$ *be continuous functions. Assume*

$$|\kappa(x, y)| \leq 1 \tag{4.8}$$

for all $x, y \in \overline{\Omega}$, *and that there exists a continuous nondecreasing function* $\psi : [0, R] \to \mathbf{R}_+$ *with* $\psi(R) > 0$, *and* $\phi \in C(\overline{\Omega}; \mathbf{R}_+)$, *such that*

$$|f(y,z)| \leq \phi(y)\,\psi(|z|) \tag{4.9}$$

for all $y \in \overline{\Omega}$, $z \in B$, *and*

$$|\phi|_{L^1(\Omega)} \leq \frac{R}{\psi(R)}. \tag{4.10}$$

Then (4.7) *has a solution* $u \in C(\overline{\Omega}; \mathbf{R}^n)$ *with* $|u|_\infty \leq R$.

Proof. The result follows from Theorem 4.2, where

$$h(x,y,z) = \kappa(x,y)\,f(y,z),$$

once we show that (4.5) holds. For this, let $u \in C(\overline{\Omega}; B)$ be any solution of (4.6) for some $\lambda \in (0,1)$. Then using (4.8)–(4.10) we obtain

$$\begin{aligned} |u(x)| &\leq \lambda \int_\Omega |\kappa(x,y)|\,\phi(y)\,\psi(|u(y)|)\,dy \\ &\leq \lambda\psi(R) \int_\Omega \phi(y)\,dy \\ &= \lambda\psi(R)\,|\phi|_{L^1(\Omega)} \\ &\leq \lambda R < R. \end{aligned}$$

Hence $|u|_\infty < R$. Therefore Theorem 4.2 applies. ■

In particular, if κ is the Green's function of a certain differential operator with respect to some boundary conditions, we can guarantee (4.5) by some other types of conditions on f.

Here are some examples for $\overline{\Omega} = [0,1]$, that is for the Hammerstein equation in $\mathbf{R}^n$

$$u(x) = \int_0^1 \kappa(x,y)\,f(y,u(y))\,dy, \quad x \in [0,1]. \tag{4.11}$$

Theorem 4.3 *Let* κ *be the Green's function of the differential operator* $-u'' + u$ *with respect to one set of the homogeneous Dirichlet, Neumann, or periodic boundary conditions*

$$u(0) = u(1) = 0 \quad (\textit{Dirichlet}), \tag{4.12}$$

$$u'(0) = u'(1) = 0 \quad (\textit{Neumann}), \tag{4.13}$$

$$u(0) - u(1) = u'(0) - u'(1) = 0 \quad (\textit{periodic conditions}). \tag{4.14}$$

Let $f : [0,1] \times B \to \mathbf{R}^n$ *be continuous. Assume that for every* $z \in B$ *with* $|z| = R$, *one has*

$$(z, f(y,z)) \leq R^2 \quad \textit{for all } y \in [0,1]. \tag{4.15}$$

Then (4.11) *has a solution* $u \in C([0,1];\mathbf{R}^n)$ *with* $|u|_\infty \leq R$.

Proof. To apply Theorem 4.2 we need to prove (4.5) for any solution $u \in C([0,1];B)$ of (4.6). Let $u \in C([0,1];B)$ be a solution of (4.6) for some $\lambda \in (0,1)$. Clearly, $|u|_\infty \leq R$. Assume

$$|u|_\infty = R.$$

Then there exists a point $x_0 \in [0,1]$ such that

$$|u(x_0)| = R.$$

On the other hand, since κ is a Green's function $u \in C^2([0,1];\mathbf{R}^n)$ and solves the problem

$$\begin{cases} -u''(x) + u(x) = \lambda f(x, u(x)), & x \in [0,1], \\ u \in \mathcal{B}. \end{cases} \tag{4.16}$$

Here $\mathcal{B}$ stands for either homogeneous Dirichlet, Neumann, or periodic boundary conditions. We shall derive a contradiction to (4.15) once we show that

$$(u(x_0), f(x_0, u(x_0))) > R^2. \tag{4.17}$$

If $x_0 \in (0,1)$ then since $|u|^2$ achieves its maximum at x_0, we must have

$$\left(|u|^2\right)''(x_0) \leq 0.$$

Direct computation shows that

$$\left(|u|^2\right)' = (u,u)' = 2(u,u')$$

and

$$\left(|u|^2\right)'' = 2|u'|^2 + 2(u,u''). \tag{4.18}$$

Consequently

$$(u(x_0), u''(x_0)) \leq 0.$$

This together with (4.16) yields

$$\begin{aligned} 0 &\geq (u(x_0), u(x_0) - \lambda f(x_0, u(x_0))) \\ &= R^2 - \lambda (u(x_0), f(x_0, u(x_0))) \\ &> \lambda \left(R^2 - (u(x_0), f(x_0, u(x_0)))\right). \end{aligned}$$

Thus (4.17) holds.

Next we prove (4.17) in case that $x_0 = 0$. This case can not hold under conditions (4.12).

Case 1. Assume $\mathcal{B}$ means (4.13). Assume

$$(u(0), f(0, u(0))) \leq R^2.$$

Then

$$\begin{aligned} R^2 &> \lambda R^2 \\ &\geq (u(0), \lambda f(0, u(0))) \\ &= (u(0), -u''(0) + u(0)) \\ &= R^2 - (u(0), u''(0)). \end{aligned}$$

It follows that $(u(0), u''(0)) > 0$. Now (4.18) implies $\left(|u|^2\right)''(0) > 0$ and so $\left(|u|^2\right)'$ is strictly increasing near 0. Then

$$\left(|u|^2\right)'(x) > \left(|u|^2\right)'(0) = 2\left(u(0), u'(0)\right) = 0$$

for $x > 0$ near 0. So $|u|^2$ is strictly increasing near 0, which is impossible since $|u|^2$ achieves its maximum at 0. Thus (4.17) holds.

Case 2. Assume $\mathcal{B}$ means (4.14). If

$$\left(u(0), u'(0)\right) = \left(u(1), u'(1)\right) \neq 0,$$

it follows that $|u|^2$ can not achieve its maximum at $x_0 = 0$ (equivalently, at 1). Hence

$$\left(u(0), u'(0)\right) = \left(u(1), u'(1)\right) = 0$$

and we make the same reasoning as at Case 1.

Finally, we prove (4.17) similarly if $x_0 = 1$.

Therefore in all cases (4.17) holds, yielding a contradiction. ∎

Remark 4.1 *Theorem 4.3 is in fact an existence result for the two-point boundary value problem*

$$\begin{cases} -u''(x) + u(x) = f(x, u(x)), & x \in [0,1] \\ u \in \mathcal{B}. \end{cases}$$

We refer the reader to Granas–Guenther–Lee [24] and O'Regan–Precup [37] for numerous results on boundary value problems via the Leray–Schauder principle.

4.3 Existence Results for Volterra Integral Equations

This section presents general existence theorems for the Volterra integral equation in $\mathbf{R}^n$

$$u(t) = \int_a^t h(t, s, u(s))\, ds, \quad t \in [a,b]. \tag{4.19}$$

We seek continuous solutions with values in the ball

$$B = \{z \in \mathbf{R}^n : |z| \leq R\},$$

i.e., $u \in C([a,b]; B)$.

First we state the analog of Theorem 4.2 for Volterra equations, another direct consequence of Theorem 4.1.

Theorem 4.4 *Let* $h : [a,b]^2 \times B \to \mathbf{R}^n$ *be continuous. Assume that*

$$|u|_\infty < R \tag{4.20}$$

for any solution $u \in C([a,b]; B)$ *to*

$$u(t) = \lambda \int_a^t h(t, s, u(s))\, ds, \quad t \in [a,b] \tag{4.21}$$

for each $\lambda \in (0,1)$. *Then* (4.19) *has a solution in* $C([a,b]; B)$.

Next we give a sufficient condition for (4.20) in the case of the equation in $\mathbf{R}^n$

$$u(t) = \int_a^t \kappa(t, s)\, f(s, u(s))\, ds, \quad t \in [a,b]. \tag{4.22}$$

Corollary 4.2 *Let $\kappa : [a,b]^2 \to \mathbf{R}$ and $f : [a,b] \times B \to \mathbf{R}^n$ be continuous functions. Assume that*

$$|\kappa(t,s)| \leq 1 \tag{4.23}$$

for all $t, s \in [a,b]$, there exists a continuous nondecreasing function $\psi : (0,R] \to (0,\infty)$ and a $\phi \in C([a,b]; \mathbf{R}_+)$ such that

$$|f(s,z)| \leq \phi(s)\,\psi(|z|) \tag{4.24}$$

for all $s \in [a,b]$, $z \in B$, and

$$|\phi|_{L^1[a,b]} \leq \int_0^R \frac{1}{\psi(\sigma)} d\sigma. \tag{4.25}$$

Then (4.22) has a solution $u \in C([a,b]; \mathbf{R}^n)$ with $|u|_\infty \leq R$.

Proof. Let $u \in C([a,b]; B)$ be any solution of (4.21) for some $\lambda \in (0,1)$. Here

$$h(t,s,z) = \kappa(t,s)\, f(s,z).$$

Then

$$|u(t)| \leq \lambda \int_a^t |\kappa(t,s)\, f(s,u(s))|\, ds \leq \lambda \int_a^t \phi(s)\,\psi(|u(s)|)\, ds \tag{4.26}$$

for all $t \in [a,b]$. Here we have understand that $\psi(0) = \lim_{t \downarrow 0} \psi(t)$. Let

$$c(t) = \min\left\{R,\ \lambda \int_a^t \phi(s)\,\psi(|u(s)|)\, ds\right\}.$$

Clearly c is nondecreasing. We claim that $c(b) < R$. Assume the contrary. Then, since $c(a) = 0$, there exists a subinterval $[a', b'] \subset [a,b]$ with

$$c(a') = 0,\ c(b') = R \text{ and } c(t) \in (0,R) \text{ for } t \subset (a', b').$$

Since by (4.26),

$$|u(t)| \leq c(t) \leq R \text{ on } [a,b],$$

and ψ is nondecreasing on $[0,R]$, we have

$$c'(s) = \lambda \phi(s)\,\psi(|u(s)|) \leq \lambda \phi(s)\,\psi(c(s))$$

for all $s \in [a', b']$. Now integration from a' to b' yields

$$\begin{aligned} \int_{a'}^{b'} \frac{c'(s)}{\psi(c(s))} ds &= \int_0^R \frac{1}{\psi(\sigma)} d\sigma \\ &\le \lambda \int_{a'}^{b'} \phi(s)\, ds \\ &\le \lambda \int_a^b \phi(s)\, ds \\ &< \int_a^b \phi(s)\, ds, \end{aligned}$$

a contradiction. Notice we may assume $|\phi|_{L^1[a,b]} > 0$ since otherwise we have nothing to prove. Hence $c(b) < R$ and so, by (4.26), $|u(t)| < R$ for all $t \in [a, b]$. Therefore $|u|_\infty < R$ and Theorem 4.4 applies. ■

Remark 4.2 Notice Corollary 4.2 can be directly derived from Schauder's fixed point theorem if we observe that $T(D) \subset D$, where

$$\begin{aligned} D &= \{u \in C([a, b]; \mathbf{R}^n) : |u(t)| \le \gamma(t) \text{ on } [a, b]\}, \\ \gamma(t) &= I^{-1}\left(\int_a^t \phi(s)\, ds\right), \\ I(t) &= \int_0^t \frac{1}{\psi(\sigma)} d\sigma, \end{aligned}$$

and T is the Volterra integral operator associated to the right hand side of the equation (4.22). We note that $\gamma(t) \le R$ for all $t \in [a, b]$ because of (4.25).

Corollary 4.2 yields a global existence result for the *initial value problem*

$$\begin{cases} u'(t) = f(t, u(t)), & t \in [0, t_1], \\ u(0) = 0. \end{cases} \tag{4.27}$$

Theorem 4.5 *Let* $f \in C([0, t_1] \times \mathbf{R}^n; \mathbf{R}^n)$. *Assume that there exists a continuous nondecreasing function* $\psi : (0, \infty) \to (0, \infty)$ *such that*

$$|f(s, z)| \le \psi(|z|)$$

for all $s \in [0, t_1]$, $z \in \mathbf{R}^n$, *and*

$$t_1 < \int_0^\infty \frac{1}{\psi(\sigma)} d\sigma.$$

Then (4.27) *has a solution* $u \in C^1([0, t_1]; \mathbf{R}^n)$.

Proof. Problem (4.27) is equivalent to (4.22). Here $[a, b] = [0, t_1]$ and $\kappa \equiv 1$. Choose any $R > 0$ such that

$$t_1 \leq \int_0^R \frac{1}{\psi(\sigma)} d\sigma.$$

Then (4.25) holds and Corollary 4.2 applies. ■

4.4 The Cauchy Problem for an Integral Equation with Delay

In this section we turn back to the initial value problem (3.9). The technique we use here is based upon the Leray–Schauder principle. In addition we seek positive solutions u with $u(t) \geq a$ for all $t \in [0, t_1]$, where $a \in \mathbf{R}_+$. Here, for $z, z' \in \mathbf{R}^n$, by $z \geq z'$ we mean $z_i \geq z'_i$, $i = 1, 2, ..., n$. Also $z \geq a$ stands for $z_i \geq a$, $i = 1, 2, ..., n$.

Our assumptions are as follows:

(i) $f \in C([-\tau, t_1] \times [a, \infty)^n; \mathbf{R}^n_+)$;
(ii) $\varphi \in C([-\tau, 0]; \mathbf{R}^n)$, satisfies (3.10) and $\varphi(t) \geq a$ on $[-\tau, 0]$;
(iii) there exists a function $g \in C([-\tau, t_1]; \mathbf{R}^n)$ such that

$$f(t, z) \geq g(t)$$

for all $t \in [-\tau, t_1]$, $z \in [a, \infty)^n$, and

$$\int_{t-\tau}^t g(s)\, ds \geq a$$

for all $t \in [0, t_1]$;

(iv) there exists a continuous nondecreasing function $\psi : (a, \infty) \to (0, \infty)$ such that

$$|f(t, z)| \leq \psi(|z|)$$

for all $t \in [0, t_1]$, $z \in [a, \infty)^n$, and

$$t_1 < \int_b^\infty \frac{1}{\psi(\sigma)} d\sigma,$$

where

$$b = \int_{-\tau}^0 |f(s, \varphi(s))|\, ds.$$

Theorem 4.6 *Assume that the assumptions (i)–(iv) are satisfied. Then (3.9) has at least one solution* $u \in C([-\tau, t_1]; \mathbf{R}^n)$ *with* $u(t) \geq a$ *for all* $t \in [-\tau, t_1]$.

Proof. We first note that $b \geq |\varphi(0)|$. Indeed,

$$b = \int_{-\tau}^{0} |f(s, \varphi(s))| \, ds \geq \left| \int_{-\tau}^{0} f(s, \varphi(s)) \, ds \right| = |\varphi(0)|.$$

We apply Theorem 4.1 with $X = C([0, t_1]; \mathbf{R}^n)$,

$$K = \{u \in X : u(0) = \varphi(0), \ u(t) \geq a \ \text{on} \ [0, t_1]\}$$

and $T : K \to K$ given by

$$T(u)(t) = \int_{t-\tau}^{t} f(s, \widetilde{u}(s)) \, ds \quad (t \in [0, t_1]),$$

where

$$\widetilde{u}(t) = \begin{cases} \varphi(t), & t \in [-\tau, 0] \\ u(t), & t \in [0, t_1]. \end{cases}$$

According to (i)–(iii), the operator T is well defined. In addition, T is completely continuous (see Theorem 3.7). In what follows we shall establish the boundedness of all solutions to

$$u = (1 - \lambda) u_0 + \lambda T(u) \tag{4.28}$$

for $\lambda \in (0, 1)$, where u_0 is the constant function $\varphi(0)$. To this end, let u be any solution of (4.28) for some $\lambda \in (0, 1)$. Then

$$|u(t)| \leq (1 - \lambda) b + \lambda \int_{t-\tau}^{t} |f(s, \widetilde{u}(s))| \, ds$$

Let $c : [0, t_1] \to \mathbf{R}$ be given by

$$c(t) = (1 - \lambda) b + \lambda \int_{t-\tau}^{t} |f(s, \widetilde{u}(s))| \, ds.$$

Notice $c(0) = b$. Furthermore, for $t \in [0, t_1]$ one has

$$\begin{aligned} c'(t) &= \lambda (|f(t, u(t))| - |f(t - \tau, \widetilde{u}(t - \tau))|) \\ &\leq \lambda |f(t, u(t))| \\ &\leq \lambda \psi(|u(t)|), \end{aligned}$$

and, since $|u(t)| \leq c(t)$ and ψ is nondecreasing,

$$c'(t) \leq \lambda \psi(c(t)).$$

Therefore

$$\frac{c'(s)}{\psi(c(s))} \leq \lambda < 1.$$

Integration from 0 to t yields

$$\int_b^{c(t)} \frac{1}{\psi(\sigma)} d\sigma = \int_0^t \frac{c'(s)}{\psi(c(s))} ds < t_1.$$

Hence $c(t) < R$, where R is any fixed number satisfying

$$t_1 \leq \int_b^R \frac{1}{\psi(\sigma)} d\sigma.$$

Therefore $|u(t)| < R$ for all $t \in [0, t_1]$. ■

4.5 Periodic Solutions of an Integral Equation with Delay

In this section we establish the existence of periodic solutions of a given period $\omega > 0$ for the integral equation in $\mathbf{R}^n$

$$u(t) = \int_{t-\tau}^{t} f(s, u(s))\, ds. \tag{4.29}$$

The ideas in this section were adapted from the paper Precup [46].

Our assumptions are:

(h1) $f \in C(\mathbf{R} \times \mathbf{R}_+^n; \mathbf{R}_+^n)$;

(h2) $f(t + \omega, z) = f(t, z)$ for every $t \in \mathbf{R}$, $z \in \mathbf{R}_+^n$;

(h3) there exists a number $a > 0$ and an ω-periodic function $g \in C(\mathbf{R}; \mathbf{R}^n)$ such that

$$f(t, z) \geq g(t)$$

for all $t \in \mathbf{R}$, $z \in [a, \infty)^n$, and

$$\int_{t-\tau}^{t} g(s)\, ds \geq a$$

for all $t \in \mathbf{R}$;

(h4) there exist two numbers b, R with

$$a\sqrt{n} < b < R,$$

and a function $\psi \in C([a,R];\mathbf{R}_+)$ with $\psi(t) > 0$ on $[b,R]$, such that

$$|f(t,z)| \leq \psi(|z|)$$

for all $t \in \mathbf{R}$ and $z \in [a,\infty)^n$ with $|z| \leq R$,

$$\omega \leq \int_b^R \frac{1}{\psi(\sigma)} d\sigma \tag{4.30}$$

and

$$|f(t,z)| \leq \frac{b}{\tau} \tag{4.31}$$

for all $t \in \mathbf{R}$ and $z \in [a,\infty)^n$ satisfying $b \leq |z| \leq R$.

Theorem 4.7 *Assume that* (h1)–(h4) *are satisfied. Then* (4.29) *has an ω-periodic continuous solution u such that*

$$u(t) \geq a \quad on \quad [0,\omega],$$

$$\min_{t\in[0,\omega]} |u(t)| < b \quad and \quad \max_{t\in[0,\omega]} |u(t)| \leq R.$$

Proof. We shall apply the Leray–Schauder principle. Here X is the space of all continuous ω-periodic functions u with the norm $|u|_\infty = \max_{t\in[0,\omega]} |u(t)|$,

$$K = \{u \in X : u(t) \geq a \text{ on } [0,\omega]\}$$

and

$$U = \left\{ u \in K : \min_{t\in[0,\omega]} |u(t)| < b,\ |u|_\infty < R \right\}.$$

Also, u_0 is the constant vector-valued function $(a, a, ..., a)$ simply denoted by a, and $T : K \to K$ is given by

$$T(u)(t) = \int_{t-\tau}^{t} f(s, u(s))\, ds \quad (t \in \mathbf{R}).$$

It is easy to show that (h1)–(h3) guarantee that T is well defined and completely continuous. We now claim that the Leray–Schauder boundary condition (4.1) is satisfied. Assume by contradiction that there are $u \in \partial U$ and $\lambda \in (0,1)$ such that

$$u = (1-\lambda)\, u_0 + \lambda T(u),$$

that is

$$u(t) = (1-\lambda)\, a + \lambda \int_{t-\tau}^{t} f(s, u(s))\, ds, \quad t \in \mathbf{R}. \tag{4.32}$$

Since $u \in \partial U$ we have either

$$|u|_\infty = R \quad \text{and} \quad \min_{t\in[0,\omega]} |u(t)| < b, \tag{4.33}$$

or

$$|u|_\infty \le R \quad \text{and} \quad \min_{t\in[0,\omega]} |u(t)| = b. \tag{4.34}$$

First assume (4.33). By differentiating (4.32) we obtain

$$u'(t) = \lambda f(t, u(t)) - \lambda f(t-\tau, u(t-\tau)).$$

Then

$$(u(t), u'(t)) = \lambda (u(t), f(t, u(t))) - \lambda (u(t), f(t-\tau, u(t-\tau))).$$

Since all the components of the vectors $u(t)$ and $f(t-\tau, u(t-\tau))$ are nonnegative, we have that

$$(u(t), f(t-\tau, u(t-\tau))) \ge 0.$$

Hence

$$(u(t), u'(t)) \le \lambda (u(t), f(t, u(t))).$$

Furthermore, using (h4) we obtain

$$(u(t), u'(t)) \le \lambda |u(t)|\, |f(t, u(t))| \le \lambda |u(t)|\, \psi(|u(t)|).$$

Since

$$|u|' = \left(\sqrt{|u|^2}\right)' = \frac{(u, u')}{|u|}$$

we obtain

$$|u(t)|' \le \lambda \psi(|u(t)|). \tag{4.35}$$

Let $t_0 \in [0,\omega]$ be such that

$$|u(t_0)| = \min_{t\in[0,\omega]} |u(t)|$$

and let $t_1 \in (t_0, t_0+\omega]$ be such that $|u(t_1)| = R$. From (4.35), by integration from t_0 to t_1, we obtain

$$\int_{|u(t_0)|}^{|u(t_1)|} \frac{1}{\psi(\sigma)} d\sigma \leq \lambda (t_1 - t_0) < \omega.$$

This, according to (4.30), is impossible since $|u(t_0)| < b$ and $|u(t_1)| = R$. Thus (4.33) can not hold.

Now assume (4.34). Let $t_0 \in [0, \omega]$ be such that

$$|u(t_0)| = \min_{t \in [0,\omega]} |u(t)| = b.$$

Then from (4.32) and (4.31) we obtain

$$b = |u(t_0)| \leq (1 - \lambda) a\sqrt{n} + \lambda b < b,$$

a contradiction.

Thus the Leray–Schauder boundary condition holds and Theorem 4.1 applies. ■

Remark 4.3 Assume (h1)–(h3) hold and that instead of (h4) , the following condition is satisfied:

(h4*) $|f(t, z)| \leq \frac{R}{\tau}$ for all $t \in [0, \omega]$ and $z \in [a, \infty)^n$ with $|z| \leq R$.

Then $T(K_R) \subset K_R$, where $K_R = \{u \in K : |u(t)| \leq R \text{ on } [0, \omega]\}$, and the existence of a continuous ω-periodic solution to (4.29) is guaranteed by Schauder's fixed point theorem.

Example 4.1 Let $n = 1$, $\tau = \omega = 1$ and let $f(t, z) = \psi_1(z)$ $(z \in \mathbf{R}_+)$, where

$$\psi_1(z) = \begin{cases} 5z, & z \in [0,1] \\ -4z + 9, & z \in [1,2] \\ 1, & z \in [2,3] \\ 3z - 8, & z \in [3,5] \\ z + 2, & z \in [5,\infty). \end{cases}$$

The assumptions (h1)–(h4) are satisfied with $a = 1$, $b = 2$, $R = 3$, $g(t) = 1$ and $\psi(\sigma) = \psi_1(\sigma)$. However, for any $R > 0$ there is no $a < R$ such that (h1)–(h3) and (h4*) are satisfied.

Remark 4.4 For a given function f satisfying (h1)–(h2) there could exist several intervals $[a, R]$ such that (h3) and (h4) hold. If these intervals are disjoint, then the corresponding solutions guaranteed by Theorem 4.7 are distinct.

Example 4.2 Let n, τ and ψ be as in Example 4.1, and let

$$f(t,z) = g_1(t)\,\psi_k(z), \quad t \in \mathbf{R},\ z \in [4(k-1), 4k],$$

$k = 1, 2, \ldots$, where

$$\psi_k(z) = 4(k-1) + \psi_1(z - 4(k-1))$$

and g_1 is any nonnegative continuous function with a period $\omega > 0$, such that

$$\int_{t-1}^{t} g_1(s)\,ds \geq 1, \quad t \in [0,\omega].$$

It is easy to see that if

$$\omega \max_{t\in[0,\omega]} g_1(t) \leq (4(k-1)+1)^{-1},$$

then all the assumptions of Theorem 4.7 are satisfied for

$$a = 4(k-1)+1, \quad b = 4(k-1)+2, \quad R = 4(k-1)+3,$$

$$\begin{aligned} g(t) &= (4(k-1)+1)\,g_1(t), \\ \psi(z) &= \psi_k(z) \max_{t\in[0,\omega]} g_1(t). \end{aligned}$$

Therefore for each $k \in N \setminus \{0\}$ satisfying

$$4(k-1)+1 \leq \left(\omega \max_{t\in[0,\omega]} g_1(t)\right)^{-1},$$

the equation (4.29) has at least one continuous ω-periodic solution u_k such that

$$4(k-1)+1 \leq \inf_{t\in\mathbf{R}} u_k(t) < 4(k-1)+2$$

and

$$\sup_{t\in\mathbf{R}} u_k(t) \leq 4(k-1)+3.$$

In particular, when $g_1(t) \equiv 1$ such solutions are the constant functions

$$u_k(t) = 4(k-1) + \frac{9}{5}$$

for $k = 1, 2, \ldots$.

Other results on the integral equation (4.29) can be found in Precup–Kirr [47] and Trif [50]. The same equation in a general Banach space is studied in Guo–Lakshmikantham–Liu [27].

For other applications of the Leray–Schauder principle to integral, differential, integro-differential and partial differential equations we refer the reader to Bobisud [5], Constantin [13], Frigon [21], Gilbarg–Trudinger [22], Granas–Guenther–Lee [24], [25], Guenther–Lee [26], Gupta [28], Lan–Webb [31], Mawhin [33], Ntouyas–Tsamatos [34], O'Regan–Meehan [35], Pachpatte [40] and Precup [45].

Chapter 5

Existence Theory in L^p Spaces

In this chapter we present three examples of continuous operators acting in L^p spaces, namely: the Nemytskii superposition operator; the Fredholm linear integral operator; and the Hammerstein nonlinear integral operator. As applications we shall prove via the Leray–Schauder principle several existence results in L^p for Hammerstein and Volterra–Hammerstein integral equations in $\mathbf{R}^n$. We show that these results immediately yield existence theorems of weak solutions (in Sobolev spaces) to the initial value and two-point boundary value problems for ordinary differential equations in $\mathbf{R}^n$, under some more general conditions than the continuity. Notice the weak solutions are functions which satisfy the differential equations almost everywhere (a.e., that is, except a set of measure zero).

In addition we show that the two cases of continuous solutions and respectively, of L^p solutions $(1 \leq p < \infty)$, can be treated together by considering $1 \leq p \leq \infty$.

5.1 The Nemytskii Operator

Let $\Omega \subset \mathbf{R}^N$ be an open set and $f : \Omega \times \mathbf{R}^m \to \mathbf{R}^n$ be a given function. The *Nemytskii operator* N_f associated to f assigns to each function $u : \Omega \to \mathbf{R}^m$, the function $N_f(u) : \Omega \to \mathbf{R}^n$, defined by

$$N_f(u)(x) = f(x, u(x)) \quad (x \in \Omega). \tag{5.1}$$

Suitable conditions on f guarantee that N_f has desired properties. Here we present such conditions in order that N_f maps $L^p(\Omega; \mathbf{R}^m)$ into $L^q(\Omega; \mathbf{R}^n)$.

Definition 5.1 One says that a function $f : \Omega \times R^m \to R^n$ satisfies the *Carathéodory conditions* if:

(*i*) $f(.,y) : \Omega \to R^n$ is measurable for every $y \in R^m$;
(*ii*) $f(x,.) : R^m \to R^n$ is continuous for a.e. $x \in \Omega$.

Lemma 5.1 *If f satisfies the Carathéodory conditions then N_f maps measurable functions into measurable functions.*

Proof. Let $u : \Omega \to \mathbf{R}^m$ be measurable. Then, there exists a sequence of finitely-valued functions (u_k) with

$$u_k(x) \to u(x) \quad \text{as } k \to \infty \text{ for a.e. } x \in \Omega.$$

This together with (ii) implies

$$N_f(u_k)(x) = f(x, u_k(x)) \to f(x, u(x)) = N_f(u)(x) \tag{5.2}$$

as $k \to \infty$ for a.e. $x \in \Omega$. On the other hand, u_k being finitely-valued admits a representation as

$$u_k(x) = \sum_{j=1}^{p_k} \chi_{kj}(x)\, y_{kj}$$

where $y_{kj} \in \mathbf{R}^m$ and χ_{kj} is the characteristic function of some subset $\Omega_{kj} \subset \Omega$, with

$$\Omega = \bigcup_{j=1}^{p_k} \Omega_{kj}, \quad \Omega_{kj} \cap \Omega_{ki} = \emptyset \quad \text{for } j \neq i.$$

It follows that

$$N_f(u_k)|_{\Omega_{kj}} = f(., y_{kj})$$

which, by (i), is a measurable function on Ω_{kj}. As a result, $N_f(u_k)$ is measurable on Ω. Now according to (5.2) the function $N_f(u)$ is measurable, as limit of a sequence of measurable functions. ■

Definition 5.2 Let $p \in [1, \infty]$ and $q \in [1, \infty)$. A function $f : \Omega \times R^m \to R^n$ is said to be (p,q)-*Carathéodory* if the following condition is satisfied:

$$\begin{cases} (a) \text{ if } 1 \leq p < \infty \text{ then } |f(x,z)| \leq g(x) + c\,|z|^{p/q} \\ \text{for a.e. } x \in \Omega, \text{all } z \in R^m \text{and some } g \in L^q(\Omega; \mathbf{R}_+), \; c \in R_+; \\ (b) \text{ if } p = \infty \text{ then for every } R > 0 \text{there is a } g_R \in L^q(\Omega) \text{ with} \\ |f(x,z)| \leq g_R(x) \text{ for a.e. } x \in \Omega \text{and all } z \in R^m \text{with } |z| \leq R. \end{cases} \tag{5.3}$$

Theorem 5.1 *Let $p \in [1,\infty]$ and $q \in [1,\infty)$. Assume that the function $f : \Omega \times \mathbf{R}^m \to \mathbf{R}^n$ is (p,q)-Carathéodory. Then the Nemytskii operator $N_f : L^p(\Omega;\mathbf{R}^m) \to L^q(\Omega;\mathbf{R}^n)$ associated to f, given by (5.1) is well defined, continuous and satisfies*

$$\begin{cases} (a) \text{ for } 1 \le p < \infty : \left|N_f(u)\right|_{L^q(\Omega;\mathbf{R}^n)} \le |g|_{L^q(\Omega)} + c\,|u|^{p/q}_{L^p(\Omega;\mathbf{R}^m)} \\ \quad \text{for all } u \in L^p(\Omega;\mathbf{R}^m); \\ (b) \text{ for } p = \infty : \left|N_f(u)\right|_{L^q(\Omega;\mathbf{R}^m)} \le |g_R|_{L^q(\Omega)} \\ \quad \text{for all } u \in L^\infty(\Omega;\mathbf{R}^m) \text{ with } |u|_\infty \le R \text{ and every } R > 0. \end{cases} \tag{5.4}$$

Here g, g_R and c are those from Definition 5.2.

Proof. *Step* 1: N_f maps $L^p(\Omega;\mathbf{R}^m)$ into $L^q(\Omega;\mathbf{R}^n)$ and satisfies (5.4). To prove this, let us consider any function $u \in L^p(\Omega;\mathbf{R}^m)$. By the definition of the space $L^p(\Omega;\mathbf{R}^m)$, u is measurable. Lemma 5.1 implies that $N_f(u)$ is measurable too. Furthermore, from (5.3), by means of the norm inequality, we obtain for $1 \le p,\ q < \infty$:

$$\begin{aligned} \left(\int_\Omega |N_f(u)(x)|^q\,dx\right)^{1/q} &\le \left\{\int_\Omega \left(|g(x)| + c\,|u(x)|^{p/q}\right)^q dx\right\}^{1/q} \\ &\le |g|_{L^q(\Omega)} + c\,|u|^{p/q}_{L^p(\Omega;\mathbf{R}^m)}. \end{aligned}$$

The case $p = \infty$ is left to the reader.

Step 2: N_f is continuous from $L^p(\Omega;\mathbf{R}^m)$ to $L^q(\Omega;\mathbf{R}^n)$. The proof is based on the well known Vitali's theorem (for a proof, see Dunford–Schwartz [20], or Pascali–Sburlan [42]):

Lemma 5.2 (Vitali) *Let (u_k) be a sequence of functions $u_k \in L^p(\Omega;\mathbf{R}^n)$ $(1 \le p < \infty)$ such that $u_k(x) \to u(x)$ as $k \to \infty$ for a.e. $x \in \Omega$. Then $u \in L^p(\Omega;\mathbf{R}^n)$ and $u_k \to u$ in $L^p(\Omega;\mathbf{R}^n)$ as $k \to \infty$ if and only if:*

(i) for each $\varepsilon > 0$, there exists a $\delta > 0$ such that

$$\int_A |u_k(x)|^p\,dx < \varepsilon$$

for all k and every measurable subset $A \subset \Omega$ with $\mu(A) < \delta$;

(ii) for each $\varepsilon > 0$, there exists a subset $D \subset \Omega$ such that $\mu(D) < \infty$ and

$$\int_{\Omega\setminus D} |u_k(x)|^p\,dx < \varepsilon$$

for all k.

Notice condition (ii) in Vitali's theorem is redundant if Ω is bounded.
Let (u_k) be any sequence of functions $u_k \in L^p(\Omega; \mathbf{R}^m)$ such that

$$u_k \longrightarrow u$$

in $L^p(\Omega; \mathbf{R}^m)$ as $k \to \infty$. Without loss of generality we may assume that $u_k(x) \to u(x)$ for a.e. $x \in \Omega$. Clearly

$$f(x, u_k(x)) \to f(x, u(x)),$$

i.e.,

$$N_f(u_k)(x) \to N_f(u)(x),$$

for a.e. $x \in \Omega$. Now, for $1 \leq p < \infty$, conditions (i) and (ii) in Vitali's theorem hold for (u_k) (the Necessity Part). Then, from (5.4), we can see that conditions (i), (ii) in Lemma 5.2, with q instead of p, also holds for $(N_f(u_k))$. Thus from Vitali's theorem (the Sufficiency Part) we obtain that

$$N_f(u_k) \to N_f(u)$$

in $L^q(\Omega; \mathbf{R}^n)$. Hence N_f is continuous. ■

Remark 5.1 The inequality in (5.4) shows that N_f is a bounded operator from $L^p(\Omega; \mathbf{R}^m)$ to $L^q(\Omega; \mathbf{R}^n)$.

5.2 The Fredholm Linear Integral Operator

Let X be a Banach space and $\Omega \subset \mathbf{R}^N$ an open set. A function $u : \Omega \to X$ is said to be *finitely-valued* if it is constant on each of a finite number of disjoint measurable sets $\Omega_j \subset \Omega$ with $\mu(\Omega_j) < \infty$ and equals zero on $\Omega \setminus \bigcup_j \Omega_j$. A function $u : \Omega \to X$ is said to *strongly measurable* if there exists a sequence of finitely-valued functions convergent in X to $u(x)$ for a.e. $x \in \Omega$. For any $p \in [1, \infty]$, we let $L^p(\Omega; X)$ be the set of all strongly measurable functions $u : \Omega \to X$ with $|u|_X \in L^p(\Omega)$.

For strongly measurable functions Egorov's theorem is still true.

Lemma 5.3 (Egorov) *Let X be a Banach space and $\Omega \subset \mathbf{R}^N$ be bounded open. If (u_k) is a sequence of strongly measurable functions from Ω to X, that strongly converges in X a.e. on Ω to a strongly measurable function $u : \Omega \to X$, then for each $\varepsilon > 0$, there exists a subset $\Omega' \subset \Omega$ such that $\mu(\Omega \setminus \Omega') < \varepsilon$ and on Ω' the convergence of $u_k(x)$ to $u(x)$ is uniform.*

Proof. First we note that if $v : \Omega \to X$ is strongly measurable then $|v|_X$ is measurable. Indeed, from the definition of a strongly measurable function it follows that there is a sequence (v_k) of finitely-valued functions such that $v_k(x) \to v(x)$ a.e. on Ω. Let

$$v_k(x) = \sum_{i=1}^{m_k} \chi_{ki}(x)\, v_{ki},$$

where $v_{ki} \in X$ and χ_{ki} is the characteristic function of a measurable subset $\Omega_{ki} \subset \Omega$, with $\Omega_{kj} \cap \Omega_{ki} = \emptyset$ for $j \neq i$ and $\Omega = \bigcup_{i=1}^{m_k} \Omega_{ki}$. Since

$$|v_k(x)|_X = \sum_{i=1}^{m_k} \chi_{ki}(x)\, |v_{ki}|_X\,,$$

we have

$$\begin{aligned} \left| |v(x)|_X - \sum_{i=1}^{m_k} \chi_{ki}(x)\, |v_{ki}|_X \right| &= ||v(x)|_X - |v_k(x)|_X| \\ &\leq |v(x) - v_k(x)|_X \to 0 \text{ as } k \to \infty. \end{aligned}$$

Hence $|v|_X$ is the limit of a sequence of measurable real functions. Thus, $|v|_X$ is measurable.

Of course we may assume that $u_k(x) \to u(x)$ for every $x \in \Omega$. By the above remark, since $u - u_k$ is strongly measurable, $|u - u_k|_X$ is measurable and so the set $\{x \in \Omega : |u(x) - u_k(x)|_X < \varepsilon\}$ is measurable. As a result the set

$$\Omega_j = \bigcap_{k=j+1}^{\infty} \{x \in \Omega : |u(x) - u_k(x)|_X < \varepsilon\}$$

is measurable too. In addition $\Omega_j \subset \Omega_i$ if $j < i$, and since $u_k(x) \to u(x)$ on Ω, one has $\Omega = \bigcup_{j=1}^{\infty} \Omega_j$. Now

$$\begin{aligned} \mu(\Omega) &= \mu(\Omega_1 \cup (\Omega_2 \setminus \Omega_1) \cup (\Omega_3 \setminus \Omega_2) \cup \ldots) \\ &= \mu(\Omega_1) + \mu(\Omega_2 \setminus \Omega_1) + \mu(\Omega_3 \setminus \Omega_2) + \ldots \\ &= \lim_{j \to \infty} \mu(\Omega_j)\,. \end{aligned}$$

Hence $\lim_{j \to \infty} \mu(\Omega \setminus \Omega_j) = 0$, and therefore for each $\eta > 0$ there is a j_0 such that

$$\mu(\Omega \setminus \Omega_j) < \eta, \quad j \geq j_0.$$

Thus for any positive integer k there exist a set $\Omega'_k \subset \Omega$ and an index N_k such that

$$\begin{aligned} \mu(\Omega'_k) &\le \frac{\varepsilon}{2^k}; \\ |u(x) - u_j(x)|_X &< \frac{1}{2^k} \quad \text{for } j > N_k,\ x \in \Omega \setminus \Omega'_k. \end{aligned}$$

Let $\Omega' = \Omega \setminus \bigcup_{k=1}^{\infty} \Omega'_k$. Then

$$\mu(\Omega \setminus \Omega') \le \sum_{k=1}^{\infty} \mu(\Omega'_k) \le \varepsilon \sum_{k=1}^{\infty} \frac{1}{2^k} = \varepsilon$$

and $u_j(x)$ converges uniformly on Ω'. ■

Using Egorov's theorem we immediately obtain that if $\Omega \subset \mathbf{R}^N$ is bounded open then for each $u \in L^p(\Omega; X)$ one has

$$\| |\tau_h(u) - u|_X |_{L^p(\Omega_h)} \to 0 \quad \text{as } h \to 0. \tag{5.5}$$

Here $\tau_h(u)$ is the translation of u by h; also recall $\Omega_h = \Omega \cap (\Omega - h)$.

For more information about the theory of functions with values in a general Banach space see Guo–Lakshmikantham–Liu [27] and Yosida [52].

Let $\kappa : \Omega^2 \to \mathbf{R}$. We say that $\kappa \in L^p(\Omega; L^r(\Omega))$ if $\kappa(x, .) \in L^r(\Omega)$ for a.e. $x \in \Omega$, and the map $x \longmapsto \kappa(x, .) \in L^r(\Omega)$ belongs to $L^p(\Omega; L^r(\Omega))$. The notation $\kappa \in C(\overline{\Omega}; L^r(\Omega))$ has a similar meaning.

Theorem 5.2 *Let $\Omega \subset \mathbf{R}^N$ be bounded open, $p, q \in [1, \infty]$, $r \in [1, \infty]$ be the conjugate of q, that is $1/q + 1/r = 1$, and let $\kappa : \Omega^2 \to \mathbf{R}$. Assume*

$$\begin{aligned} &(a) \ \text{if } 1 \le p < \infty \text{ then } \kappa \in L^p(\Omega; L^r(\Omega)); \\ &(b) \ \text{if } p = \infty \text{ then } \kappa \in C(\overline{\Omega}; L^r(\Omega)). \end{aligned} \tag{5.6}$$

Then the Fredholm linear integral operator $A : L^q(\Omega; \mathbf{R}^n) \to L^p(\Omega; \mathbf{R}^n)$ *with kernel κ given by*

$$A(u)(x) = \int_\Omega \kappa(x, y)\, u(y)\, dy \quad (x \in \Omega)$$

is well defined and completely continuous. Moreover, for $p = \infty$ we have

$$A(L^q(\Omega; \mathbf{R}^n)) \subset C(\overline{\Omega}; \mathbf{R}^n).$$

Proof. 1) If $u \in L^q(\Omega; \mathbf{R}^n)$ then $A(u)$ is a measurable function. We omit the details.

2) A is well defined and bounded (hence continuous). Indeed, if $u \in L^q(\Omega; \mathbf{R}^n)$ then by Hölder's inequality we have

$$|A(u)(x)| \le \left(\int_\Omega |\kappa(x,y)|^r\, dy\right)^{1/r} \left(\int_\Omega |u(y)|^q\, dy\right)^{1/q}.$$

Hence

$$\left(\int_\Omega |A(u)(x)|^p\, dx\right)^{1/p} \le |u|_q \left\{\int_\Omega \left(\int_\Omega |\kappa(x,y)|^r\, dy\right)^{p/r} dx\right\}^{1/p},$$

that is

$$|A(u)|_p \le \| \kappa(x,.)|_r|_p\, |u|_q\,.$$

This shows that A is well defined and bounded. Of course, some changes have to be made in the proof if $r = \infty$, $q = \infty$, or $p = \infty$.

3) Let $p = \infty$ and let $u \in L^q(\Omega; \mathbf{R}^n)$. We show that $A(u) \in C(\overline{\Omega}; \mathbf{R}^n)$. Indeed, for every $x, x' \in \overline{\Omega}$, we have

$$\begin{aligned} |A(u)(x) - A(u)(x')| &\le \int_\Omega |\kappa(x,y) - \kappa(x',y)|\, |u(y)|\, dy \\ &\le |u|_q\, |\kappa(x,.) - \kappa(x',.)|_r \to 0 \end{aligned}$$

as $|x - x'| \to 0$. Hence $A(u)$ is a continuous function.

4) A is completely continuous. To prove this, let $X \subset L^q(\Omega; \mathbf{R}^n)$ be a bounded set. Assume that

$$|u|_q \le m \quad \text{for all} \quad u \in X.$$

We have to show that $Y := A(X)$ is relatively compact in $L^p(\Omega; \mathbf{R}^n)$. For this we use Theorem 1.4. Firstly,

$$|A(u)(x)| \le \int_\Omega |\kappa(x,y)|\, |u(y)|\, dy \le m\, |k(x,.)|_r =: \nu(x)$$

for a.e. $x \in \Omega$, and $\nu \in L^p(\Omega)$, so (1.4) holds. Secondly,

$$\begin{aligned} |\tau_h(A(u))(x) - A(u)(x)| &\le \int_\Omega |\tau_h^1(\kappa)(x,y) - \kappa(x,y)|\, |u(y)|\, dy \\ &\le m\, |\tau_h^1(\kappa)(x,.) - \kappa(x,.)|_r\,. \end{aligned}$$

Here $\tau_h^1(\kappa)$ is the *translation* of κ by h with respect to the first variable, i.e.,

$$\tau_h^1(\kappa)(x,y) = \begin{cases} \kappa(x+h,y) & \text{if } x+h \in \Omega, \\ 0 & \text{if } x+h \notin \Omega. \end{cases}$$

It follows that

$$|\tau_h(A(u)) - A(u)|_{L^p(\Omega_h;\mathbf{R}^n)} \leq m \left| |\tau_h^1(\kappa)(x,.) - \kappa(x,.)|_r \right|_{L^p(\Omega_h)}.$$

But from (5.5) we have

$$\left| |\tau_h^1(\kappa)(x,.) - \kappa(x,.)|_r \right|_{L^p(\Omega_h)} \to 0 \quad \text{as } h \to 0.$$

Hence (1.5) also holds. Consequently Y is relatively compact in $L^p(\Omega;\mathbf{R}^n)$. ■

Remark 5.2 Notice if $\kappa : \Omega^2 \to R$ is measurable, $\kappa(x,.) \in L^r(\Omega)$ for a.e. $x \in \Omega$, and the map $x \longmapsto |\kappa(x,.)|_r$ belongs to $L^p(\Omega)$, then the map $x \longmapsto \kappa(x,.) \in L^r(\Omega)$ is strongly measurable, and so $\kappa \in L^p(\Omega; L^r(\Omega))$.

5.3 The Hammerstein Integral Operator

Consider the *Hammerstein integral operator* given by

$$T(u)(x) = \int_\Omega \kappa(x,y) f(y,u(y))\, dy \quad (x \in \Omega). \tag{5.7}$$

This operator appears as the composition of the Fredholm linear integral operator A of kernel κ with the Nemytskii operator N_f associated to f, that is

$$T = AN_f.$$

Theorem 5.3 *Let $\Omega \subset \mathbf{R}^N$ be open, $\kappa : \Omega^2 \to \mathbf{R}$ and $f : \Omega \times \mathbf{R}^n \to \mathbf{R}^n$. Let $p \in [1,\infty]$, $q \in [1,\infty)$ and let $r \in (1,\infty]$ be the conjugate of q. Assume that the Fredholm linear integral operator $A : L^q(\Omega;\mathbf{R}^n) \to L^p(\Omega;\mathbf{R}^n)$ of kernel κ is well defined and completely continuous. In addition assume that f is a (p,q)-Carathéodory function. Then the Hammerstein integral operator $T : L^p(\Omega;\mathbf{R}^n) \to L^p(\Omega;\mathbf{R}^n)$ given by* (5.7) *is well defined and completely continuous.*

Proof. From Theorem 5.1, the Nemytskii operator N_f is well defined, bounded and continuous from $L^p(\Omega;\mathbf{R}^n)$ into $L^q(\Omega;\mathbf{R}^n)$. Now the conclusion follows from Theorem 2.1, since $T = AN_f$. ■

Remark 5.3 According to Theorem 5.2, if Ω is bounded and (5.6) holds for κ, then A is well defined and completely continuous; moreover, in this case, for $p = \infty$,

$$T\left(L^{\infty}\left(\Omega;\mathbf{R}^n\right)\right) \subset C\left(\overline{\Omega};\mathbf{R}^n\right),$$

and so any fixed point of T belongs to $C\left(\overline{\Omega};\mathbf{R}^n\right)$.

5.4 Hammerstein Integral Equations

This section presents a very general existence principle for both L^p solutions and continuous solutions to the Hammerstein integral equation in $\mathbf{R}^n$

$$u(x) = \int_{\Omega} \kappa(x,y)\, f(y,u(y))\, dy \quad \text{a.e. on } \Omega. \tag{5.8}$$

Here $\Omega \subset \mathbf{R}^N$ is an open set. For a given $p \in [1,\infty]$, we seek a weak solution, that is a function $u \in L^p(\Omega;\mathbf{R}^n)$ which satisfies (5.8) for almost every $x \in \Omega$. In addition, we look for solutions in $\overline{U}$, where U is a bounded open set of $L^p(\Omega;\mathbf{R}^n)$ containing the null function.

Theorem 5.4 *Let $\Omega \subset \mathbf{R}^N$ be open, $\kappa : \Omega^2 \to \mathbf{R}$ and $f : \Omega \times \mathbf{R}^n \to \mathbf{R}^n$. Assume that there exist $p \in [1,\infty]$, $q \in [1,\infty)$ such that the Fredholm linear integral operator $A : L^q(\Omega;\mathbf{R}^n) \to L^p(\Omega;\mathbf{R}^n)$ with kernel κ is well defined and completely continuous and f is (p,q)-Carathéodory. In addition assume that there exists a bounded open set $U \subset L^p(\Omega;\mathbf{R}^n)$ containing the null function, such that*

$$u \in U \tag{5.9}$$

for any solution $u \in \overline{U}$ to

$$u(x) = \lambda \int_{\Omega} \kappa(x,y)\, f(y,u(y))\, dy \quad \text{a.e. on } \Omega, \tag{5.10}$$

for each $\lambda \in (0,1)$. Then (5.8) has a solution in $L^p(\Omega;\mathbf{R}^n)$ with $u \in \overline{U}$.

Proof. Apply Theorem 4.1 to $K = X = L^p(\Omega;\mathbf{R}^n)$ with norm $|.|_p$, u_0 the null function, and $T : \overline{U} \to L^p(\Omega;\mathbf{R}^n)$ given by

$$T(u)(x) = \int_{\Omega} \kappa(x,y)\, f(y,u(y))\, dy \quad (x \in \Omega).$$

This operator is completely continuous by Theorem 5.3. ■

Remark 5.4 If Ω is bounded and (5.6) holds for κ, then the Fredholm linear integral operator A of kernel κ is well defined and completely continuous from $L^q(\Omega; \mathbf{R}^n)$ to $L^p(\Omega; \mathbf{R}^n)$; in this case, for $p = \infty$ the solutions of (5.8) are continuous functions on $\overline{\Omega}$.

Corollary 5.1 *Let $\Omega \subset \mathbf{R}^N$ be a bounded open set, $\kappa : \Omega^2 \to \mathbf{R}$ and $f : \Omega \times \mathbf{R}^n \to \mathbf{R}^n$. Assume that there exists $p \in [1, \infty]$, $q \in [1, \infty)$ such that $\kappa \in L^p(\Omega; L^r(\Omega))$ $(1/r + 1/q = 1)$, f is (p,q)-Carathéodory and*

$$\begin{cases} (a) \text{ if } 1 \le p < \infty \text{ then } \left(|g|_q + cR^{p/q}\right) \left||\kappa(x,.)|_r\right|_p \le R; \\ (b) \text{ if } p = \infty \text{ then } |g_R|_q \left||\kappa(x,.)|_r\right|_\infty \le R. \end{cases} \tag{5.11}$$

Here g, g_R and c are those from Definition 5.2. In addition, if $p = \infty$ assume that the Fredholm linear integral operator $A : L^q(\Omega; \mathbf{R}^n) \to L^\infty(\Omega; \mathbf{R}^n)$ with kernel κ is completely continuous. Then (5.8) *has a solution $u \in L^p(\Omega; \mathbf{R}^n)$ with $|u|_p \le R$.*

Proof. Let

$$U = \left\{ u \in L^p(\Omega; \mathbf{R}^n) : |u|_p < R \right\}$$

and $B = \overline{U}$. Let $u \in B$ be any solution of (5.10) for some $\lambda \in (0, 1)$. Then

$$\begin{aligned} |u(x)| &\le \lambda \int_\Omega |\kappa(x,y)| \, |f(y, u(y))| \, dy \\ &\le \lambda \int_\Omega |\kappa(x,y)| \left(g(y) + c\,|u(y)|^{p/q} \right) dy \\ &\le \lambda \left(\int_\Omega |\kappa(x,y)|^r \, dy \right)^{1/r} \left(\int_\Omega \left(g(y) + c\,|u(y)|^{p/q} \right)^q dy \right)^{1/q} \\ &\le \lambda \left(|g|_q + c\,|u|_p^{p/q} \right) |\kappa(x,.)|_r . \end{aligned}$$

It follows that

$$|u|_p \le \lambda \left(|g|_q + c\,|u|_p^{p/q} \right) \left||\kappa(x,.)|_r\right|_p . \tag{5.12}$$

Since $|u|_p \le R$ and $\lambda < 1$, (5.11) and (5.12) guarantee $|u|_p < R$, i.e., (5.9). Similar estimations can be obtained if $p = \infty$. ■

We note that in fact (5.11) implies $T(B) \subset B$. So Corollary 5.1 can be directly derived from Schauder's fixed point theorem.

As in Section 4.2, a better result holds when κ is the Green's function of a certain differential operator, with respect to some boundary conditions.

For instance, for the equation in $\mathbf{R}^n$

$$u(x) = \int_0^1 \kappa(x,y) f(y, u(y))\, dy \quad \text{a.e. on } (0,1), \tag{5.13}$$

the following existence theorem is true.

Theorem 5.5 *Let κ be the Green's function of the differential operator $-u'' + u$ with respect to one set of the homogeneous Dirichlet, Neumann, or periodic boundary conditions*

$$u(0) = u(1) = 0 \quad (\textit{Dirichlet}),$$

$$u'(0) = u'(1) = 0 \quad (\textit{Neumann}),$$

$$u(0) - u(1) = u'(0) - u'(1) = 0 \quad (\textit{periodic conditions}).$$

Assume that $f : (0,1) \times \mathbf{R}^n \to \mathbf{R}^n$ is a (∞, q)-Carathéodory function for some $q \in [1, \infty)$. In addition, assume that there exists $R > 0$ such that for every $z \in \mathbf{R}^n$ with $|z| = R$, one has

$$(z, f(y,z)) \le R^2 \quad \textit{for a.e. } y \in (0,1).$$

Then (5.13) *has a solution $u \in W^{2,q}(0,1;\mathbf{R}^n)$ with $|u|_\infty \le R$.*

For any real number $1 \le q \le \infty$ the *Sobolev spaces* $W^{m,q}(a,b;\mathbf{R}^n)$ $(m \in \mathbf{N} \setminus \{0\})$ are inductively defined as follows. A function u belongs to $W^{1,q}(a,b;\mathbf{R}^n)$ if it is continuous on $[a,b]$ and there exists a function $v \in L^q(a,b;\mathbf{R}^n)$ such that

$$u(t) = u(a) + \int_a^t v(s)\, ds, \quad t \in [a,b].$$

It is clear that if $u \in W^{1,q}(a,b;\mathbf{R}^n)$ then u is absolutely continuous on $[a,b]$ (note that any absolutely continuous function $u : [a,b] \to \mathbf{R}^n$ is differentiable almost everywhere on $[a,b]$ and $u' \in L^1(a,b;\mathbf{R}^n)$), $u' \in L^q(a,b;\mathbf{R}^n)$, and

$$u(t) = u(a) + \int_a^t u'(s)\, ds, \quad t \in [a,b].$$

Furthermore, for any integer $m > 1$

$$u \in W^{m,q}(a,b;\mathbf{R}^n) \quad \text{if} \quad u,\, u' \in W^{m-1,q}(a,b;\mathbf{R}^n).$$

If $u \in W^{m,q}(a,b;\mathbf{R}^n)$, then $u \in C^{m-1}([a,b];\mathbf{R}^n)$, $u^{(m-1)}$ is absolutely continuous on $[a,b]$,

$$u^{(m)} := \left(u^{(m-1)}\right)' \in L^q(a,b;\mathbf{R}^n),$$

and

$$u^{(m-1)}(t) = u^{(m-1)}(a) + \int_a^t u^{(m)}(s)\,ds, \quad t \in [a,b].$$

The space $W^{m,q}(a,b;\mathbf{R}^n)$ is a Banach space with the norm

$$|u|_{m,q} = \max\left\{\left|u^{(j)}\right|_q; \quad j = 0, 1, ..., m\right\}.$$

For more information about the theory of Sobolev spaces we refer the reader to Brezis [6].

5.5 Volterra–Hammerstein Integral Equations

We conclude this section by some results on equation (5.8) in the Volterra case. More exactly, we consider the *Volterra–Hammerstein equation* in $\mathbf{R}^n$

$$u(t) = \int_a^t \kappa(t,s)\, f(s,u(s))\,ds \quad \text{a.e. on } (a,b). \tag{5.14}$$

This equation can viewed as a special case of equation (5.8), where $\Omega = (a,b)$ and

$$\kappa(t,s) = 0 \quad \text{for } t < s.$$

The existence of solutions to (5.14) can be established without condition (5.11).

Theorem 5.6 *Let* (a,b) *be a bounded real interval,* $\kappa : (a,b)^2 \to \mathbf{R}$ *and* $f : (a,b) \times \mathbf{R}^n \to \mathbf{R}^n$. *Assume that there exists* $p \in [1,\infty)$ *and* $q \in [p,\infty) \cap (1,\infty)$ *such that* $\kappa \in L^p(a,b;L^r(a,b))$ $(1/q + 1/r = 1)$ *and* f *is* (p,q)-*Carathéodory. In addition assume that if* $q = p$, *then there exists an* $r' > r$ *such that*

$$\kappa(t,.) \in L^{r'}(a,b) \;\; \text{for a.e. } t \in (a,b), \text{ and}$$
$$\text{the map } t \longmapsto |\kappa(t,.)|_{r'} \text{ belongs to } L^p(a,b).$$

Then (5.14) *has a solution* $u \in L^p(a,b;\mathbf{R}^n)$.

Proof. First assume $p < q$. Then there exists $R > 0$ with

$$\left(|g|_q + cR^{p/q}\right) \left|\left|\kappa\left(t, .\right)\right|_r\right|_p \leq R$$

and the conclusion follows from Corollary 5.1.

Next, assume $p = q > 1$. We shall apply Theorem 5.4 to

$$U = \{u \in L^p\left(a, b; \mathbf{R}^n\right) : \|u\| < R\},$$

any number $R > 0$ and a suitable equivalent norm $\|.\|$ on $L^p\left(a, b; \mathbf{R}^n\right)$. Let $u \in L^p\left(a, b; \mathbf{R}^n\right)$ be any solution of

$$u\left(t\right) = \lambda \int_a^t \kappa\left(t, s\right) f\left(s, u\left(s\right)\right) ds \quad \text{a.e. on} \quad \left(a, b\right), \tag{5.15}$$

for some $\lambda \in (0, 1)$. For any $\theta > 0$ we have

$$\begin{aligned} |u\left(t\right)| &\leq \lambda \int_a^t |\kappa\left(t, s\right)| \left(g\left(s\right) + c\left|u\left(s\right)\right|\right) ds \\ &\leq \lambda \int_a^t |\kappa\left(t, s\right)| e^{\theta(s-a)} \left(g\left(s\right) + c\left|u\left(s\right)\right| e^{-\theta(s-a)}\right) ds. \end{aligned}$$

Define an equivalent norm $\|.\|$ on $L^p\left(a, b; \mathbf{R}^n\right)$, by

$$\|u\| = \left(\int_a^b \left(\left|u\left(s\right)\right| e^{-\theta(s-a)}\right)^p ds\right)^{1/p}.$$

Since

$$\frac{1}{r'} + \frac{r' - r}{rr'} + \frac{1}{p} = 1,$$

Hölder's inequality applied to three functions yields

$$\begin{aligned} |u\left(t\right)| &\leq \lambda \left(\int_a^b |\kappa\left(t, s\right)|^{r'} ds\right)^{1/r'} \left(\int_a^t e^{\theta rr'(s-a)/(r'-r)} ds\right)^{(r'-r)/(rr')} \\ &\quad \times \left(\int_a^b \left(g\left(s\right) + c\left|u\left(s\right)\right| e^{-\theta(s-a)}\right)^p ds\right)^{1/p} \\ &\leq \lambda \left(|g|_p + c\|u\|\right) \left(\frac{r' - r}{\theta rr'}\right)^{(r'-r)/(rr')} e^{\theta(t-a)} \left|\kappa\left(t, .\right)\right|_{r'}. \end{aligned}$$

It follows that

$$\|u\| \leq \lambda \left(|g|_p + c\|u\|\right) \left(\frac{r' - r}{\theta rr'}\right)^{(r'-r)/(rr')} \left|\left|\kappa\left(t, .\right)\right|_{r'}\right|_p.$$

Now we choose $\theta > 0$ such that

$$\left(|g|_p + cR\right)\left(\frac{r'-r}{\theta r r'}\right)^{(r'-r)/(rr')} \left||\kappa(t,.)|_{r'}\right|_p \leq R$$

Then, since $\lambda < 1$, $\|u\| < R$ and Theorem 5.4 applies. ∎

For $p = \infty$ a better result can be established.

Theorem 5.7 *Let (a,b) be a bounded real interval, $\kappa : (a,b)^2 \to \mathbf{R}$ and $f : (a,b) \times \mathbf{R}^n \to \mathbf{R}^n$. Assume that f satisfies the Carathéodory conditions, there exist $\alpha, \beta \in (1,\infty]$, $\gamma \in [1,\infty)$ with $\alpha' < \beta$ $(1/\alpha + 1/\alpha' = 1)$ and $1/\alpha + 1/\beta + 1/\gamma = 1$, $\phi \in L^\alpha(a,b)$, and a continuous nondecreasing function $\psi : (0,\infty) \to (0,\infty)$ such that*

$$\kappa \in L^\infty\left(a,b; L^\beta(a,b)\right),$$

$$|f(s,z)| \leq \phi(s)\,\psi(|z|^\gamma),$$

for a.e. $s \in (a,b)$ and all $z \in \mathbf{R}^n$, and

$$\left[\left||k(t,.)|_\beta\right|_\infty |\phi|_\alpha\right]^\gamma (b-a) \leq \int_0^\infty \frac{1}{[\psi(\sigma)]^\gamma}\, d\sigma. \tag{5.16}$$

In addition assume that the Fredholm linear integral operator A with kernel κ is is well defined and completely continuous from $L^{\beta'}(a,b;\mathbf{R}^n)$ to $C([a,b];\mathbf{R}^n)$ $(1/\beta + 1/\beta' = 1)$. Then (5.14) has a solution $u \in C([a,b];\mathbf{R}^n)$.

Proof. We shall apply Theorem 5.4 with $p = \infty$, any finite q satisfying $\beta' < q \leq \alpha$ $(1/\beta + 1/\beta' = 1)$, and

$$U = \{u \in L^\infty(a,b;\mathbf{R}^n) : |u|_\infty < R\}$$

with a suitable radius $R > 0$. Let us show that such an R can be found so that condition (5.9) holds.

Let $u \in L^\infty(a,b;\mathbf{R}^n)$ be any solution of (5.15) for some $\lambda \in (0,1)$. Then $u \in C([a,b];\mathbf{R}^n)$ since A has values in $C([a,b];\mathbf{R}^n)$. Using Hölder's inequality, we obtain

$$\begin{aligned} |u(t)| &\leq \lambda \int_0^t |\kappa(t,s)|\,\phi(s)\,\psi(|u(s)|^\gamma)\, ds \\ &\leq \lambda\, |\kappa(t,.)|_\beta\, |\phi|_\alpha \left(\int_a^t \psi(|u(s)|^\gamma)^\gamma\, ds\right)^{1/\gamma}. \end{aligned}$$

Hence

$$|u(t)|^\gamma \le c(t),$$

where

$$c(t) = \left[\lambda \left|\left|\kappa(t,.)\right|_\beta\right|_\infty |\phi|_\alpha\right]^\gamma \left(\int_a^t \psi\left(\left|u(s)\right|^\gamma\right)^\gamma ds\right).$$

From

$$\begin{aligned} c'(s) &= \left[\lambda \left|\left|\kappa(t,.)\right|_\beta\right|_\infty |\phi|_\alpha\right]^\gamma \psi\left(\left|u(s)\right|^\gamma\right)^\gamma \\ &\le \left[\lambda \left|\left|\kappa(t,.)\right|_\beta\right|_\infty |\phi|_\alpha\right]^\gamma \psi(c(s))^\gamma, \end{aligned}$$

by integration from a to t, we obtain

$$\begin{aligned} \int_0^{c(t)} \frac{1}{\psi(\sigma)^\gamma} d\sigma &= \int_a^t \frac{c'(s)}{\psi(c(s))^\gamma} ds \\ &\le \left[\lambda \left|\left|\kappa(t,.)\right|_\beta\right|_\infty |\phi|_\alpha\right]^\gamma (b-a) \\ &< \left[\left|\left|\kappa(t,.)\right|_\beta\right|_\infty |\phi|_\alpha\right]^\gamma (b-a). \end{aligned}$$

This together with (5.16) shows that there exists an $R > 0$ independent of λ, with

$$c(t) < R^\gamma \quad \text{for all } t \in [a,b].$$

Thus $|u(t)| < R$ on $[a,b]$. ■

As a consequence we have the following global existence result for the Cauchy problem

$$\begin{cases} u' = f(t,u) \quad \text{a.e. on } (a,b), \\ u(a) = 0. \end{cases} \tag{5.17}$$

Theorem 5.8 *Let (a,b) be a bounded real interval and $f : (a,b) \times \mathbf{R}^n \to \mathbf{R}^n$. Assume that f satisfies the Carathéodory conditions, there exists $\alpha \in (1,\infty]$, $\gamma \in [1,\infty)$ with $1/\alpha + 1/\gamma = 1$, $\phi \in L^\alpha(a,b)$, and a nondecreasing continuous function $\psi : (0\infty) \to (0,\infty)$ such that*

$$|f(s,z)| \le \phi(s)\,\psi\left(|z|^\gamma\right)$$

for a.e. $s \in (a,b)$ and all $z \in \mathbf{R}^n$, and

$$|\phi|_\alpha^\gamma (b-a) \le \int_0^\infty \frac{1}{[\psi(\sigma)]^\gamma} d\sigma. \tag{5.18}$$

Then (5.14) has a solution $u \in W^{1,\alpha}(a,b;\mathbf{R}^n)$.

Proof. Apply Theorem 5.7 with $\beta = \infty$ and $\kappa(t,s) = 1$ for $s \leq t$, $\kappa(t,s) = 0$ for $t < s$. Notice the Fredholm linear integral operator A is now given by

$$A(u)(t) = \int_a^t u(s)\,ds,$$

and is completely continuous from $L^1(a,b;\mathbf{R}^n)$ to $C([a,b];\mathbf{R}^n)$. ∎

The results in Sections 5.4–5.5 are presented here for the first time. We have adapted here some ideas from the papers of O'Regan–Precup [36], [38], [39]. For similar results and extensions to equations and inclusions in Banach spaces, see Couchouron–Precup [17].

For other methods, results and applications in nonlinear integral and integro-differential equations, see the papers Appell–DePascale–Nguyêñ–Zabreiko [2], Appell–Vignoli–Zabreiko [3], Brezis–Browder [7], Browder [9], Bugajewski–Szufla [10], Burton [12], Couchouron–Kamenski [16], Papageorgiou [41], Petruşel [43], and the monographs Agarwal–O'Regan [1], Barbu [4], Burton [11], Corduneanu [14], [15], Deimling [18], Guo–Lakshmikantham-Liu [27], Krasnoselskii [30], Rus [48] and Väth [51].

References: Part I

[1] AGARWAL, R.P. and O'REGAN, D., *Infinite Interval Problems for Differential, Difference and Integral Equations*, Kluwer Academic Publishers, Dordrecht, 2001.

[2] APPELL, J., De PASCALE, E., NGUYÊÑ, H.T. and ZABREIKO, P.P., Nonlinear integral inclusions of Hammerstein type, *Topol. Methods Nonlinear Anal.* **5** (1995), 111–124.

[3] APPELL, J., VIGNOLI, A. and ZABREIKO, P.P., Implicit function theorems and nonlinear integral equations, *Exposition. Math.* **14** (1996), 385–424.

[4] BARBU, V., *Nonlinear Semigroups and Differential Equations in Banach Spaces,* Ed. Acad.–Noordhoff, Bucureşti–Leyden, 1976.

[5] BOBISUD, L.E., Existence of positive solutions to some nonlinear singular boundary value problems on finite and infinite intervals, *J. Math. Anal. Appl.* **173** (1993), 69–83.

[6] BREZIS, H., *Analyse fonctionnelle,* Masson, Paris, 1983.

[7] BREZIS, H. and BROWDER, F.E., Nonlinear integral equations and systems of Hammerstein type, *Adv. Math.* **18** (1975), 115–144.

[8] BROUWER, L.E.J., Über abbildungen von mannigfaltigkeiten, *Math. Ann.* **71** (1912), 97–115.

[9] BROWDER, F.E., Nonlinear functional analysis and nonlinear integral equations of Hammerstein and Fredholm type. In: *Contributions to Nonlinear Functional Analysis,* Academic Press, 1971, 425–500.

[10] BUGAJEWSKI, D. and SZUFLA, S., Existence theorems for differential and integral equations in locally convex spaces, *Discussiones Math.* **13** (1993), 141–147.

[11] BURTON, T.A., *Volterra Integral and Differential Equations,* Academic Press, New York, 1983.

[12] BURTON, T.A., Basic neutral integral equations of advanced type, *Nonlinear Anal.* **31** (1998), 295–310.

[13] CONSTANTIN, A., Topological transversality: application to an integrodifferential equation, *J. Math. Anal. Appl.* **197** (1996), 855–863.

[14] CORDUNEANU, C., *Integral Equations and Stability of Feedback Systems,* Academic Press, New York, 1973.

[15] CORDUNEANU, C., *Integral Equations and Applications,* Cambridge Univ. Press, New York, 1991.

[16] COUCHOURON, J.-F. and KAMENSKI, M., A unified topological point of view for integro-differential inclusions. In: *Differential Inclusions and Optimal Control* (J. Andres, L. Górniewicz and P. Nistri eds.), Lectures Notes in Nonlinear Anal. Vol. 2, 1998, 123–137.

[17] COUCHOURON, J.-F. and PRECUP, R., Existence principles for inclusions of Hammerstein type involving noncompact acyclic multivalued maps, *Electron. J. Differential Equations* **2002** (2002), No. 04, 1–21.

[18] DEIMLING, K., *Nonlinear Functional Analysis,* Springer–Verlag, Berlin–Heidelberg–New York, 1985.

[19] DUGUNDJI, J. and GRANAS, A., *Fixed Point Theory,* Monografie Matematyczne, PWN, Warsaw, 1982.

[20] DUNFORD, N. and SCHWARTZ, J.T., *Linear Operators, Part* I, Interscience Publishers, New York–London, 1958.

[21] FRIGON, M., Application de la théorie de la transversalité topologique à des problèmes non linéaires pour des équations différentielles ordinaires, *Dissertationes Math.* **296**, 1990.

[22] GILBARG, D. and TRUDINGER, N.S., *Elliptic Partial Differential Equations of Second Order,* Springer–Verlag, Berlin–Heidelberg–New York–Tokyo, 1983.

[23] GRANAS, A., Homotopy extension theorem in Banach spaces and some of its applications to the theory of non-linear equations, *Bull. Acad. Pol. Sci.* **7** (1959), 387–394.

[24] GRANAS, A., GUENTHER, R. and LEE, J., Nonlinear boundary value problems for ordinary differential equations, *Dissertationes Math.* **244,** 1985.

[25] GRANAS, A., GUENTHER, R. and LEE, J., Some general existence principles in the Carathéodory theory of nonlinear differential systems, *J. Math. Pures Appl.* **70** (1991), 153–196.

[26] GUENTHER, R.B. and LEE, J.W., Some existence results for nonlinear integral equations via topological transversality, *J. Integral Equations Appl.* **3** (1993), 195–209.

[27] GUO, D., LAKSHMIKANTHAM, V. and LIU, X., *Nonlinear Integral Equations in Abstract Spaces,* Kluwer Academic Publishers, Dordrecht–Boston–London, 1996.

[28] GUPTA, C.P., Existence and uniqueness theorems for a fourth order boundary value problem of Sturm–Liouville type, *Differential Integral Equations* **4** (1991), 397–410.

[29] KANTOROVITCH, L. and AKILOV, G., *Analyse fonctionnelle,* Mir Publishers, Moscow, 1981.

[30] KRASNOSELSKII, M.A., *Topological Methods in the Theory of Nonlinear Integral Equations,* Pergamon Press, Oxford–London–New York–Paris, 1964.

[31] LAN, K. and WEBB, J.R.L., Positive solutions of semilinear differential equations with singularities, *J. Differential Equations* **148** (1998), 407–421.

[32] LERAY, J. and SCHAUDER, J., Topologie et équations fonctionnelles, *Ann. Sci. École Norm. Sup.* **51** (1934), 45–78.

[33] MAWHIN, J., Leray–Schauder continuation theorems in the absence of a priori bounds, *Topol. Methods Nonlinear Anal.* **9** (1997), 179–200.

[34] NTOUYAS, S.K. and TSAMATOS, P.CH., Global existence for functional integro-differential equations of delay and neutral type, *Appl. Anal.* **54** (1994), 251–262.

[35] O'REGAN, D. and MEEHAN, M., *Existence Theory for Nonlinear Integral and Integrodifferential Equations,* Kluwer Academic Publishers, Dordrecht–Boston–London, 1998.

[36] O'REGAN, D. and PRECUP, R., Fixed point theorems for set-valued maps and existence principles for integral inclusions, *J. Math. Anal. Appl.* **245** (2000), 594–612.

[37] O'REGAN, D. and PRECUP, R., *Theorems of Leray–Schauder Type and Applications,* Gordon and Breach Science Publishers, Amsterdam, 2001.

[38] O'REGAN, D. and PRECUP, R., Existence criteria for integral equations in Banach spaces, *J. Inequal. Appl.* **6** (2001), 77–97.

[39] O'REGAN, D. and PRECUP, R., Integrable solutions of Hammerstein integral inclusions in Banach spaces, *Dynam. Contin. Discrete Impuls. Systems* **9** (2002), 165–176.

[40] PACHPATTE, B.G., Application of the Leray–Schauder alternative to some Volterra integral and integrodifferential equations, *Indian J. Pure Appl. Math.* **26** (1995), 1161–1168.

[41] PAPAGEORGIOU, N.S., Existence of solutions for integrodifferential inclusions in Banach spaces, *Comment. Math. Univ. Carolin.* **32** (1991), 687–696.

[42] PASCALI, D. and SBURLAN, S., *Nonlinear Mappings of Monotone Type,* Ed. Acad.–Sijthoff & Noordhoff, Bucureşti–Alphen aan den Rijn, 1978.

[43] PETRUŞEL, A., Integral inclusions. Fixed point approaches, *Ann. Soc. Math. Pol.* **40** (2000), 147–158.

[44] PRECUP, R., Positive solutions of the initial value problem for an integral equation modeling infectious disease. In: *Seminar on Fixed Point Theory* (I.A. Rus ed.), Babeş–Bolyai University, Faculty of Mathematics, Research Seminars, Preprint nr. 3, Cluj, 1991, 25–30.

[45] PRECUP, R., *Nonlinear Integral Equations* (Romanian), Babeş–Bolyai University, Cluj, 1993.

[46] PRECUP, R., Periodic solutions for an integral equation from biomathematics via Leray–Schauder principle, *Studia Univ. Babeş–Bolyai Math.* **39** (1994), no.1, 47–58.

[47] PRECUP, R. and KIRR, E., Analysis of a nonlinear integral equation modelling infection diseases. In: *Proc. Internat. Conf. Timişoara, May 19–21, 1997* (St. Balint ed.), Univ. of Timişoara, 1997, 178–195.

[48] RUS, I.A., *Principles and Applications of the Fixed Point Theory* (Romanian), Dacia, Cluj–Napoca, 1979.

[49] SCHAUDER, J., Der fixpunktsatz in functionalräumen, *Studia Math.* **2** (1930), 171–180.

[50] TRIF, T., Positive solutions of a nonlinear integral equation from biomathematics, *Demonstratio Math.* **32** (1999), 129–138.

[51] VÄTH, M., *Volterra and Integral Equations of Vector Functions,* Marcel Dekker, New York–Basel, 2000.

[52] YOSIDA, K., *Functional Analysis,* Springer–Verlag, Berlin, 1974.

[53] ZEIDLER, E., *Nonlinear Functional Analysis and Its Applications,* Vol I, Springer–Verlag, New York–Berlin–Heidelberg–Tokyo, 1986.

Part II

VARIATIONAL METHODS

Chapter 6

Positive Self-Adjoint Operators in Hilbert Spaces

In this part we present some variational methods with applications to the existence of L^p solutions of the Hammerstein integral equation in $\mathbf{R}^n$

$$u(x) = \int_\Omega \kappa(x,y)\, f(y, u(y))\, dy \quad \text{a.e. on } \Omega. \tag{6.1}$$

We show that this equation can be put under the *variational form*

$$E'(u) = 0.$$

The construction of the functional E associated to equation (6.1) is based on the theory of positive self-adjoint linear operators in Hilbert spaces.

The purpose of this introductory chapter is to present some definitions and results of this theory that will be required in what follows. We define the notions of adjoint operator, self-adjoint operator, positive operator, and square root of a positive self-adjoint linear operator. As an application we shall deal with bounded linear operators $A : L^q(\Omega; \mathbf{R}^n) \to L^p(\Omega; \mathbf{R}^n)$, where $1/p + 1/q = 1$. The goal is to split such an operator in the form HH^*, where $H : L^2(\Omega; \mathbf{R}^n) \to L^p(\Omega; \mathbf{R}^n)$ and H^* is the adjoint of H. When A is the Fredholm linear integral operator of kernel κ, such a representation is useful in order to transfer the operator equation equivalent to (6.1) from the Banach space $L^p(\Omega; \mathbf{R}^n)$ to the Hilbert space $L^2(\Omega; \mathbf{R}^n)$.

6.1 Adjoint Operators

Let X be a Hilbert space, Y a Banach space, and Y^* the dual of Y. We shall denote by $(.,.)$ both inner product of X and *duality* between Y and

Y^*, i.e., the bilinear functional on $Y^* \times Y$, defined as $(v^*, v) = v^*(v)$ for all $v \in Y$ and $v^* \in Y^*$.

Proposition 6.1 *For each bounded linear operator $A : X \to Y$ there exists a unique bounded linear operator $A^* : Y^* \to X$ such that*

$$(v^*, A(u)) = (A^*(v^*), u) \tag{6.2}$$

for all $u \in X$ and $v^* \in Y^*$.

Proof. For any fixed $v^* \in Y^*$ define a bounded linear functional $f : X \to \mathbf{R}$, by $f(u) = (v^*, A(u))$. According to Riesz' representation theorem there exists a unique $w \in X$ such that $f(u) = (w, u)$ for all $u \in X$. Define $A^*(v^*) = w$. From (6.2), since A is linear and continuous, it immediately follows that A^* is a continuous linear operator. ∎

Definition 6.1 The operator A^* given by Proposition 6.1 is called the *adjoint operator* of A.

Definition 6.2 A bounded linear operator $A : X \to X$ is said to be *self-adjoint* if $A = A^*$, i.e.,

$$(v, A(u)) = (A(v), u)$$

for all $u, v \in X$.

Proposition 6.2 *If $A : X \to X$ is a self-adjoint operator, then*

$$|A| = \sup\{|(A(u), u)| : |u| = 1\}. \tag{6.3}$$

Proof. By definition, the norm of the operator A is given by

$$|A| = \sup\{|A(u)| : |u| \le 1\}.$$

For any $u \in X$ with $|u| \le 1$, we have

$$|(A(u), u)| \le |A(u)|\,|u| \le |A|.$$

Then, if we denote $\gamma = \sup\{|(A(u), u)| : |u| = 1\}$, we have $\gamma \le |A|$. To prove the converse inequality we first note that for every $u \in X$ one has

$$|(A(u), u)| \le \gamma\,|u|^2.$$

Then since

$$(A(u+v), u+v) - (A(u-v), u-v) = 4(A(u), v)$$

we obtain

$$|(A(u), v)| \leq \frac{\gamma}{4}\left(|u+v|^2 + |u-v|^2\right) = \frac{\gamma}{2}\left(|u|^2 + |v|^2\right).$$

If we choose

$$v = \frac{|u|}{|A(u)|} A(u) \quad (\text{for } A(u) \neq 0),$$

we deduce that $|A(u)| \leq \gamma |u|$, which is also true if $A(u) = 0$. Hence $|A| \leq \gamma$, and so $|A| = \gamma$. ■

Definition 6.3 A linear operator $A : X \to X$ is said to be *positive* $(A \geq 0)$ if $(A(u), u) \geq 0$ for all $u \in X$.

Proposition 6.3 *If* $A : X \to X$ *is a positive self-adjoint operator, then*

$$|(A(u), v)|^2 \leq (A(u), u)(A(v), v)$$

for all u, $v \in X$.

Proof. A being positive, one has $(A(u+tv), u+tv) \geq 0$, that is

$$t^2 (A(v), v) + t[(A(u), v) + (A(v), u)] + (A(u), u) \geq 0$$

for all $t \in \mathbf{R}$. Since A is self-adjoint this yields

$$t^2 (A(v), v) + 2t(A(u), v) + (A(u), u) \geq 0$$

for all $t \in \mathbf{R}$. Consequently

$$(A(u), v)^2 - (A(u), u)(A(v), v) \leq 0$$

and the proof is complete. ■

If A and B are two linear operators from X into X we say that $A \leq B$ if $B - A$ is a positive operator. From (6.3) we deduce that if A, B are self-adjoint then

$$0 \leq A \leq B \quad \text{implies} \quad |A| \leq |B|.$$

Proposition 6.4 *Let* (A_k) *be a sequence of positive self-adjoint operators* $A_k : X \to X$ *and let* $B : X \to X$ *be a bounded linear operator. Assume that*

$$A_k \leq A_{k+1} \leq B$$

for all $k \in \mathbf{N}$. *Then for every* $u \in X$ *the sequence* $(A_k(u))$ *is convergent to some element* $A(u)$, *and the operator* $A : X \to X$ *defined in this way is positive, self-adjoint, and satisfies*

$$A_k \leq A \leq B, \quad k \in \mathbf{N}.$$

Proof. For every $k \in \mathbf{N}$, we have

$$0 \leq (A_k(u), u) \leq (A_{k+1}(u), u) \leq (B(u), u).$$

Hence the sequence of real numbers $(A_k(u), u)$ is convergent. We shall prove that $(A_k(u))$ is a Cauchy sequence. Indeed, for $i < j$ one has $A_j - A_i \geq 0$, and from Proposition 6.3 we obtain

$$\begin{aligned} |(A_j - A_i)(u)|^4 &= ((A_j - A_i)(u), (A_j - A_i)(u))^2 \\ &\leq ((A_j - A_i)(u), u)\left((A_j - A_i)^2(u), (A_j - A_i)(u)\right) \\ &\leq |(A_j - A_i)(u)|^2 |A_j - A_i| \, [(A_j(u), u) - (A_i(u), u)]. \end{aligned}$$

From $0 \leq A_j - A_i \leq B$, we have $|A_j - A_i| \leq |B|$. Then

$$|A_j(u) - A_i(u)|^2 \leq |B| \, [(A_j(u), u) - (A_i(u), u)].$$

Hence the sequence $(A_k(u))$ is Cauchy and so convergent. The rest of the proof is left to the reader. ■

6.2 The Square Root of a Positive Self-Adjoint Operator

Definition 6.4 Let $A : X \to X$ be a positive self-adjoint operator. A positive self-adjoint operator $B : X \to X$ is said to be the *square root* of A if $B^2 = A$.

Proposition 6.5 *For each positive self-adjoint operator* A *the square root* B *exists and is unique. Moreover, if* A *is completely continuous* B *is completely continuous too.*

Proof. 1) *Existence.* We may assume without loss of generality that $|A| = 1$. Then $A \leq I$, where I is the identity map of X. We define the operators $B_k : X \to X$ for $k \geq 2$ by $B_2 \equiv 0$, and

$$B_{k+1} = B_k + \frac{1}{2}\left(A - B_k^2\right), \tag{6.4}$$

for $k = 2, 3, \ldots$. We easily check that B_k is self-adjoint,

$$\begin{aligned} B_j B_i &= B_i B_j, \\ I - B_{k+1} &= \frac{1}{2}(I - B_k)^2 + \frac{1}{2}(I - A) \end{aligned}$$

and

$$B_{k+1} - B_k = \frac{1}{2}\left[(I - B_{k-1}) + (I - B_k)\right](B_k - B_{k-1}).$$

It follows that

$$B_k \leq I, \quad B_k \leq B_{k+1}$$

for all $k \in \{2, 3, \ldots\}$. In particular, $B_k \geq 0$ since $B_2 = 0$. By Proposition 6.4, there exists a positive self-adjoint operator B with

$$B(u) = \lim_{k \to \infty} B_k(u), \quad u \in X.$$

Letting $k \to \infty$ in (6.4), we obtain

$$B = B + \frac{1}{2}\left(A - B^2\right),$$

whence $B^2 = A$.

2) *Uniqueness.* First note if B_0 is positive, self-adjoint, and $B_0^2 = A$, then $AB_0 = B_0 A$. By induction, using (6.4) we obtain that $B_k B_0 = B_0 B_k$, whence letting $k \to \infty$, $BB_0 = B_0 B$. Now let $u \in X$ and $v = (B - B_0)(u)$. We have

$$\begin{aligned} (B(v), v) + (B_0(v), v) &= ((B + B_0)(v), v) \\ &= ((B + B_0)(B - B_0)(u), v) \\ &= \left(\left(B^2 - B_0^2\right)(u), v\right) \\ &= 0. \end{aligned}$$

Since B, $B_0 \geq 0$, we may infer that

$$(B(v), v) = (B_0(v), v) = 0.$$

On the other hand, since B is positive and self-adjoint there exists a positive self-adjoint operator B_1 with $B = B_1^2$. From

$$|B_1(v)|^2 = \left(B_1^2(v), v\right) = (B(v), v) = 0,$$

it follows that $B_1(v) = 0$ and so $B(v) = B_1(B_1(v)) = 0$. Similarly $B_0(v) = 0$. Hence

$$|B(u) - B_0(u)|^2 = \left((B - B_0)^2(u), u\right) = ((B - B_0)(v), u) = 0.$$

Thus $B(u) = B_0(u)$ for any $u \in X$. Therefore $B = B_0$.

We shall denote by $A^{1/2}$ the square root of A.

3) *Complete continuity.* Assume A is completely continuous and let (u_k) be any bounded sequence of elements of X. Then there exists a subsequence (u_{k_j}) of (u_k) such that $(A(u_{k_j}))$ is convergent. Using the identity

$$\left|A^{1/2}(u) - A^{1/2}(v)\right|^2 = (u - v, A(u) - A(v)),$$

we deduce that the sequence $(A^{1/2}(u_{k_j}))$ is Cauchy, hence convergent. Hence $A^{1/2}$ is completely continuous. ■

6.3 Splitting of Linear Operators in L^p Spaces

In this section we establish the possibility of representing certain linear operators $A : L^q(\Omega; \mathbf{R}^n) \to L^p(\Omega; \mathbf{R}^n)$ $(1/p + 1/q = 1)$ in the form

$$A = HH^*,$$

where $H : L^2(\Omega; \mathbf{R}^n) \to L^p(\Omega; \mathbf{R}^n)$ and H^* is the adjoint of H.

The presentation in this section was adapted from the book of Krasnoselskii [19].

Assume $\Omega \subset \mathbf{R}^N$ is a bounded open set. Recall that for $2 \le p < p_0$, one has

$$L^{p_0}(\Omega; \mathbf{R}^n) \subset L^p(\Omega; \mathbf{R}^n) \subset L^2(\Omega; \mathbf{R}^n) \subset L^q(\Omega; \mathbf{R}^n) \subset L^{q_0}(\Omega; \mathbf{R}^n)$$

where q, q_0 are the conjugate exponents of p and p_0, respectively.

Theorem 6.1 *Let* $2 < p_0 < \infty$, $1/p_0 + 1/q_0 = 1$ *and* $A : L^{q_0}(\Omega; \mathbf{R}^n) \to L^{p_0}(\Omega; \mathbf{R}^n)$ *a bounded linear operator. Assume that the restriction of* A *to* $L^2(\Omega; \mathbf{R}^n)$, *i.e.,* $A : L^2(\Omega; \mathbf{R}^n) \to L^2(\Omega; \mathbf{R}^n)$ *is a positive self-adjoint* *operator. Then for each* $p \in [2, p_0)$ *the operator* $A^{1/2} : L^2(\Omega; \mathbf{R}^n) \to L^2(\Omega; \mathbf{R}^n)$ *satisfies*

$$A^{1/2}\left(L^2(\Omega; \mathbf{R}^n)\right) \subset L^p(\Omega; \mathbf{R}^n)$$

and the operator

$$H : L^2(\Omega; \mathbf{R}^n) \to L^p(\Omega; \mathbf{R}^n), \quad H(u) = A^{1/2}(u),$$

is bounded (equivalently, continuous). If in addition, $A : L^2(\Omega; \mathbf{R}^n) \to L^2(\Omega; \mathbf{R}^n)$ *is completely continuous then* H *is also completely continuous.*

Proof. 1) Let $u \in L^2(\Omega; \mathbf{R}^n)$. Let us consider the sets

$$\begin{aligned} \Omega_0 &= \left\{x \in \Omega : \left|A^{1/2}(u)(x)\right| \le 1\right\}; \\ \Omega_j &= \left\{x \in \Omega : 2^{j-1} < \left|A^{1/2}(u)(x)\right| \le 2^j\right\} \end{aligned}$$

for $j = 1, 2, \ldots$. Let $w^j : \Omega \to \mathbf{R}^n$ be given by

$$w_i^j(x) = \begin{cases} \operatorname{sign}\left(A^{1/2}(u)(x)\right)_i & \text{if } x \in \Omega_j, \\ 0 & \text{if } x \notin \Omega_j, \end{cases}$$

$i = 1, 2, \ldots, n$. We have

$$\begin{aligned} 2^{j-1}\mu(\Omega_j) &\le \int_\Omega \left(A^{1/2}(u)(x), w^j(x)\right) dx \\ &= \left(A^{1/2}(u), w^j\right)_2 \\ &= \left(u, A^{1/2}(w^j)\right)_2 \\ &\le |u|_2 \left|A^{1/2}(w^j)\right|_2 \\ &= |u|_2 \left(A(w^j), w^j\right)_2^{1/2}. \end{aligned}$$

Also

$$\left(A(w^j), w^j\right)_2 \le \left|A(w^j)\right|_{p_0} \left|w^j\right|_{q_0} \le |A|_0 \left|w^j\right|_{q_0}^2 \le n\,|A|_0\, \mu(\Omega_j)^{2/q_0}.$$

Here $|A|_0$ stands for the norm of A as an operator from $L^{q_0}(\Omega; \mathbf{R}^n)$ to $L^{p_0}(\Omega; \mathbf{R}^n)$. It follows that

$$2^{j-1}\mu(\Omega_j) \le \sqrt{n}\,|u|_2\,|A|_0^{1/2}\,\mu(\Omega_j)^{1/q_0}$$

whence

$$\mu(\Omega_j) \le c\,2^{-j\,p_0}, \quad \text{where} \quad c = \left(2\sqrt{n}\,|u|_2\,|A|_0^{1/2}\right)^{p_0}.$$

Then

$$\int_{\Omega_j} \left|A^{1/2}(u)(x)\right|^p dx \le c\,2^{j\,(p-p_0)}$$

$(j = 1, 2, \ldots)$. For $j = 0$, since $p \ge 2$ one has

$$\begin{aligned}
\int_{\Omega_0} \left|A^{1/2}(u)(x)\right|^p dx &\le \int_{\Omega_0} \left|A^{1/2}(u)(x)\right|^2 dx \\
&\le \int_{\Omega} \left|A^{1/2}(u)(x)\right|^2 dx \\
&= (A(u), u)_2 \\
&\le |A|\,|u|_2^2.
\end{aligned}$$

Consequently

$$\begin{aligned}
\int_{\Omega} \left|A^{1/2}(u)(x)\right|^p dx &= \sum_{j=0}^{\infty} \int_{\Omega_j} \left|A^{1/2}(u)(x)\right|^p dx \\
&\le |A|\,|u|_2^2 + c\sum_{j=1}^{\infty} 2^{j(p-p_0)} \\
&= |A|\,|u|_2^2 + c\frac{2^{p-p_0}}{1-2^{p-p_0}} \\
&= |A|\,|u|_2^2 + \frac{c}{2^{p_0-p}-1} \\
&< \infty.
\end{aligned}$$

Hence $A^{1/2}(u) \in L^p(\Omega; \mathbf{R}^n)$ and

$$|H(u)|_p \le \left(|A|\,|u|_2^2 + \frac{c}{2^{p_0-p}-1}\right)^{1/p},$$

which shows that H is a bounded operator.

2) Assume A is completely continuous as operator from $L^2(\Omega; \mathbf{R}^n)$ to $L^2(\Omega; \mathbf{R}^n)$. Then by Proposition 6.5 $A^{1/2}$ is also completely continuous from $L^2(\Omega; \mathbf{R}^n)$ to $L^2(\Omega; \mathbf{R}^n)$. The complete continuity of H now follows from the next lemma.

Lemma 6.1 *Let $p > 2$ and let $Y \subset L^p(\Omega; \mathbf{R}^n)$ be relatively compact in $L^2(\Omega; \mathbf{R}^n)$. If for each $\varepsilon > 0$ there exists a $\delta_\varepsilon > 0$ such that*

$$\int_D |u(x)|^p\,dx < \varepsilon^p \tag{6.5}$$

for all $u \in Y$ *and* $D \subset \Omega$ *with* $\mu(D) < \delta_\varepsilon$, *then* Y *is relatively compact in* $L^p(\Omega; \mathbf{R}^n)$.

Indeed, if $Z \subset L^2(\Omega; \mathbf{R}^n)$ is any bounded set then $Y = H(Z) \subset L^p(\Omega; \mathbf{R}^n)$, is relatively compact in $L^2(\Omega; \mathbf{R}^n)$, and for every $u \in Z$ we have

$$\begin{aligned} \int_D |H(u)(x)|^p \, dx &= \int_{D \cap (\Omega_0 \cup \Omega_1 \cup \ldots \cup \Omega_{j_0})} |H(u)(x)|^p \, dx \\ &\quad + \sum_{j=j_0+1}^{\infty} \int_{D \cap \Omega_j} |H(u)(x)|^p \, dx \\ &\le 2^{j_0 p} \mu(D) + c_0 \sum_{j=j_0+1}^{\infty} 2^{j(p-p_0)}, \end{aligned}$$

where

$$c_0 = \sup\left\{ \left(2\sqrt{n} \, |u|_2 \, |A|_0^{1/2} \right)^{p_0} : u \in Z \right\}.$$

For j_0 large enough we have

$$\sum_{j=j_0+1}^{\infty} 2^{j(p-p_0)} \le \frac{\varepsilon^p}{2c_0}$$

and setting $\delta_\varepsilon = \varepsilon^p \, 2^{-1-j_0 p}$, we see that Y satisfies (6.5). Hence by Lemma 6.1 $H(Z)$ is relatively compact in $L^p(\Omega; \mathbf{R}^n)$ and the proof of Theorem 6.1 is complete. ■

Proof of Lemma 6.1. By Hausdorff's theorem it sufficies to show that for each $\varepsilon > 0$ there exists in $L^p(\Omega; \mathbf{R}^n)$ a relatively compact ε-net for Y. To this end, for each $a > 0$, we consider the set of functions $Y_a = \{u_a : u \in Y\}$, where

$$u_a(x) = \begin{cases} u(x) & \text{if } |u(x)| < a, \\ a \, |u(x)|^{-1} u(x) & \text{if } |u(x)| \ge a. \end{cases}$$

a) For each $\varepsilon > 0$ there exists a sufficiently large $a > 0$ such that Y_a is an ε-net in $L^p(\Omega; \mathbf{R}^n)$ for Y. Indeed, let δ_ε be such that (6.5) holds for all $u \in Y$ and $D \subset \Omega$ with $\mu(D) < \delta_\varepsilon$. For any $u \in Y$ we let

$$\Omega_u = \{x \in \Omega : |u(x)| \ge a\}.$$

Then, if $\mu(\Omega_u) < \delta_\varepsilon$, we have

$$\begin{aligned} |u - u_a|_p &= \left(\int_{\Omega_u} \left| u(x) - \frac{a}{|u(x)|} u(x) \right|^p dx \right)^{1/p} \\ &\leq \left(\int_{\Omega_u} |u(x)|^p dx \right)^{1/p} \\ &< \varepsilon. \end{aligned} \tag{6.6}$$

Let r be such that $|u|_2 \leq r$ for all $u \in Y$. Then

$$a\,\mu(\Omega_u)^{1/2} \leq \left(\int_{\Omega_u} |u(x)|^2 dx \right)^{1/2} \leq \left(\int_{\Omega} |u(x)|^2 dx \right)^{1/2} \leq r.$$

Hence $\mu(\Omega_u) \leq (r/a)^2$. Now choose $a > 0$ such that $(r/a)^2 < \delta_\varepsilon$. Then $\mu(\Omega_u) < \delta_\varepsilon$, and so (6.6) holds.

b) Now we show that Y_a is relatively compact in $L^p(\Omega; \mathbf{R}^n)$. Since Y is assumed to be relatively compact in $L^2(\Omega; \mathbf{R}^n)$, for a given $\varepsilon > 0$ there exists in $L^2(\Omega; \mathbf{R}^n)$ a finite ε-net for Y, say $\{u^1, u^2, ..., u^m\}$. We claim that the set

$$\{u_a^1, u_a^2, ..., u_a^m\}$$

is a $(2a)^{1-2/p}\varepsilon^{2/p}$-net in $L^p(\Omega; \mathbf{R}^n)$ for Y_a. Indeed,

$$\begin{aligned} \left|u_a - u_a^j\right|_p &= 2a \left\{ \int_\Omega \left(\frac{1}{2a} \left|u_a - u_a^j\right| \right)^p dx \right\}^{1/p} \\ &\leq 2a \left\{ \int_\Omega \left(\frac{1}{2a} \left|u - u^j\right| \right)^2 dx \right\}^{1/p} \\ &< (2a)^{1-2/p} \varepsilon^{2/p}, \end{aligned}$$

for any $j \in \{1, 2, ..., m\}$ with $\left|u - u^j\right|_2 < \varepsilon$. This completes the proof of Lemma 6.1. ■

Theorem 6.2 *Under the assumptions of Theorem 6.1, for each* $p \in [2, p_0)$, *the operator* $A : L^q(\Omega; \mathbf{R}^n) \to L^p(\Omega; \mathbf{R}^n)$ $(1/p + 1/q = 1)$ *can be represented in the form* $A = HH^*$, *where* $H : L^2(\Omega; \mathbf{R}^n) \to L^p(\Omega; \mathbf{R}^n)$ *is a bounded linear operator and* H^* *is the adjoint of* H.

Proof. Observe that the restriction of H^* to $L^2(\Omega; \mathbf{R}^n)$ coincides with $A^{1/2}$. Hence

$$HH^*|_{L^2(\Omega;\mathbf{R}^n)} = A^{1/2}A^{1/2} = A|_{L^2(\Omega;\mathbf{R}^n)}.$$

Finally, use the density of $L^2(\Omega;\mathbf{R}^n)$ in $L^q(\Omega;\mathbf{R}^n)$. ■

Now let us return to the Hammerstein integral equation (6.1) and assume that the the operators

$$N_f : L^p(\Omega;\mathbf{R}^n) \to L^q(\Omega;\mathbf{R}^n), \quad N_f(u)(x) = f(x, u(x))$$

and

$$A : L^q(\Omega;\mathbf{R}^n) \to L^p(\Omega;\mathbf{R}^n), \quad A(u)(x) = \int_\Omega \kappa(x,y)\,u(y)\,dy,$$

where $1/p + 1/q = 1$, are well defined and A admits a representation in the form $A = HH^*$, where $H : L^2(\Omega;\mathbf{R}^n) \to L^p(\Omega;\mathbf{R}^n)$ is a bounded linear operator and H^* is the adjoint of H.

Solving (6.1) in $L^p(\Omega;\mathbf{R}^n)$ is equivalent to solving the operator equation

$$u = AN_f(u), \quad u \in L^p(\Omega;\mathbf{R}^n). \tag{6.7}$$

The next proposition allows us to replace the operator equation (6.7) in the Banach space $L^p(\Omega;\mathbf{R}^n)$ by the operator equation

$$v = H^*N_fH(v), \quad v \in L^2(\Omega;\mathbf{R}^n) \tag{6.8}$$

in the Hilbert space $L^2(\Omega;\mathbf{R}^n)$.

Proposition 6.6 *If $u \in L^p(\Omega;\mathbf{R}^n)$ solves* (6.7) *then $v = H^*N_f(u)$ is a solution of* (6.8). *Conversely, if $v \in L^2(\Omega;\mathbf{R}^n)$ solves* (6.8) *then $u = H(v)$ is a solution of* (6.7). *Moreover, there is a one-to-one correspondence between the solutions of* (6.7) *and the solutions of* (6.8).

Proof. 1) Assume that $u \in L^p(\Omega;\mathbf{R}^n)$ solves (6.7) and let $v = H^*N_f(u)$. Then

$$H^*N_fH(v) = H^*N_fHH^*N_f(u) = H^*N_fAN_f(u) = H^*N_f(u) = v.$$

Hence v is a solution of (6.8).

2) Assume now that $v \in L^2(\Omega;\mathbf{R}^n)$ solves (6.8) and let $u = H(v)$. Then

$$AN_f(u) = AN_fH(v) = HH^*N_fH(v) = H(v) = u.$$

Hence u is a solution of (6.7).

3) Let $\mathcal{S}_1$, $\mathcal{S}_2$ be the sets of all solutions of (6.7) and (6.8), respectively. For any $u \in \mathcal{S}_1$, one has

$$HH^*N_f(u) = AN_f(u) = u,$$

and for any $v \in \mathcal{S}_2$,

$$H^* N_f H(v) = v.$$

These show that the map $H : \mathcal{S}_2 \to \mathcal{S}_1$ is invertible and its inverse is $H^* N_f$. Thus there is a one-to-one correspondence between the solutions of (6.7) and the solutions of (6.8) . ■

In what follows we deal with the operator equation (6.8) in $L^2(\Omega; \mathbf{R}^n)$. The main idea is to build a functional $E : L^2(\Omega; \mathbf{R}^n) \to \mathbf{R}$ whose Fréchet derivative E' is $I - H^* N_f H$. Thus the equation (6.8) can be written in the variational form

$$E'(v) = 0$$

and its solutions appear as critical points of E. By a *variational method* for (6.8) we mean any technique for proving the existence of critical points of E.

Chapter 7

The Fréchet Derivative and Critical Points of Extremum

In this chapter we present the notion of Fréchet derivative of a functional and we illustrate it by some examples. Then we build a functional $E : L^2(\Omega; \mathbf{R}^n) \to \mathbf{R}$ whose Fréchet derivative is the operator

$$I - H^* N_f H : L^2(\Omega; \mathbf{R}^n) \to L^2(\Omega; \mathbf{R}^n)$$

associated to (6.8). We prove the infinite-dimensional version of the classical Fermat's theorem about the connection between extremum points and critical points, and we give sufficient conditions for that a functional admits minimizers. The abstract results are then applied to establish the existence of L^p solutions for Hammerstein integral equations in $\mathbf{R}^n$.

7.1 The Fréchet Derivative. Examples

Let X be a Banach space, $U \subset X$ an open set, $E : U \to \mathbf{R}$ a functional, and $u \in U$ a given point.

Definition 7.1 1) The *derivative of* E *in direction* $v \in X$ at u is defined as

$$\lim_{t \to 0^+} t^{-1} (E(u + tv) - E(u))$$

if this limit exists.

2) E is said to be *Gâteaux differentiable* at u if there exists an $E'(u) \in X^*$ such that

$$(E'(u), v) = \lim_{t \to 0^+} t^{-1} (E(u + tv) - E(u))$$

for all $v \in X$. The element $E'(u)$ is called the *Gâteaux derivative* of E at u.

3) E is said to be *Fréchet differentiable* at u if there exists an $E'(u) \in X^*$ such that

$$E(u+v) - E(u) = (E'(u), v) + \omega(u, v) \tag{7.1}$$

and

$$\omega(u, v) = o(|v|), \text{ i.e, } \frac{\omega(u,v)}{|v|} \to 0, \tag{7.2}$$

as $v \to 0$. The element $E'(u)$ is called the *Fréchet derivative* of E at u.

Remark 7.1 More generally, one can define the *Fréchet derivative* at u of a map E from an open subset containing u of a Banach space X to the Banach space Y as a bounded linear operator $E'(u)$ from X to Y for which (7.1) and (7.2) hold as $v \to 0$.

We note that if E is Fréchet differentiable at a point u then E is continuous at u. This immediately follows since

$$|E(u+v) - E(u)| \leq |E'(u)| |v| + |\omega(u,v)| \leq (|E'(u)| + 1) |v| \tag{7.3}$$

for $|v| < \delta$. Here $\delta > 0$ is such that (7.1) and

$$\frac{|\omega(u,v)|}{|v|} \leq 1$$

hold for all $v \in X$ with $|v| < \delta$.

We say that the functional E is Gâteaux (Fréchet) differentiable in a subset D of U, if E is Gâteaux (respectively, Fréchet) differentiable at any point $u \in D$.

Proposition 7.1 *If E is Gâteaux differentiable in an open neighborhood V of u and $E' : V \to X^*$ is continuous at u, then E is Fréchet differentiable at u and the two derivatives at u coincide.*

Proof. Let $r \in (0, 1]$ be such that $B_r(u) \subset V$. Since $E' : V \to X^*$ is continuous at u, for every $\varepsilon > 0$ there exists a $\delta \in (0, r]$ such that

$$|E'(u+tv) - E'(u)| < \varepsilon \quad \text{for } |v| < \delta, \ |t| \leq 1. \tag{7.4}$$

On the other hand, for each $v \in B_r(0)$ the function

$$t \in [0, 1] \longmapsto (E'(u+tv), v)$$

is the derivative of the function

$$g(t) = E(u + tv) \quad (t \in [0,1]).$$

Consequently

$$\int_0^1 (E'(u + tv), v)\, dt = g(1) - g(0) = E(u + v) - E(u).$$

Hence

$$E(u + v) - E(u) - (E'(u), v) = \int_0^1 (E'(u + tv) - E'(u), v)\, dt. \qquad (7.5)$$

Let

$$\omega(u, v) = \int_0^1 (E'(u + tv) - E'(u), v)\, dt.$$

From (7.4), for all $v \in B_\delta(0)$ we have

$$\begin{aligned} |\omega(u, v)| &\leq \int_0^1 |E'(u + tv) - E'(u)|\, |v|\, dt \\ &< \varepsilon |v|. \end{aligned}$$

Hence $\omega(u, v) = o(|v|)$ as $v \to 0$. Then, (7.5) shows that $E'(u)$ is the Fréchet derivative of E at u. ■

We say that $E \in C^1(U)$ if E is Fréchet differentiable in U and its Fréchet derivative $E' : U \to X^*$ is continuous. Equivalently, in view of Proposition 7.1 $E \in C^1(U)$ if E is Gâteaux differentiable in U and its Gâteaux derivative $E' : U \to X^*$ is continuous.

We say that $E \in C^1(\overline{U})$ if $E \in C^1(U)$ and E, E' can be extended continuously to $\overline{U}$.

Example 7.1 Let $X = R^N$, $\Omega \subset R^N$ open, and $E : \Omega \to R$. If E is differentiable in a neighborhood of a point $x \in \Omega$ in the classical sense, and its partial derivatives $\partial E / \partial x_i$, $i = 1, 2, ..., N$, are continuous at x, then E is Fréchet differentiable at x and its Fréchet derivative coincides with its *gradient*, that is

$$E'(x) = \nabla E(x) = \left(\frac{\partial E}{\partial x_1}(x), \frac{\partial E}{\partial x_2}(x), ..., \frac{\partial E}{\partial x_N}(x) \right).$$

Example 7.2 Assume X is a Hilbert space and

$$E(u) = \frac{1}{2}|u|^2 \quad (u \in X).$$

Then $E \in C^1(X)$ and

$$\left(E'(u), v\right) = (u, v), \quad v \in X.$$

Indeed, for every $u, v \in X$ we have

$$E(u + tv) - E(u) = \frac{1}{2}|u + tv|^2 - \frac{1}{2}|u|^2 = t(u, v) + \frac{t^2}{2}|v|^2.$$

Hence

$$\lim_{t \to 0+} t^{-1}(E(u + tv) - E(u)) = (u, v).$$

Example 7.3 Let X be a Hilbert space, Y a Banach space, $H : X \to Y$ a bounded linear operator, and $J : Y \to R$ Fréchet differentiable in Y. Then the functional $JH : X \to R$ is Fréchet differentiable in X and

$$(JH)' = H^* J' H.$$

Here again H^* is the adjoint of H. Indeed, for every $u, v \in X$, since J is Fréchet differentiable in Y, one has

$$\begin{aligned} JH(u + v) - JH(u) &= J(H(u) + H(v)) - J(H(u)) \\ &= \left(J'(H(u)), H(v)\right) + \omega(H(u), H(v)) \\ &= \left(H^* J' H(u), v\right) + \omega(H(u), H(v)) \end{aligned}$$

and

$$\frac{\omega(H(u), H(v))}{|H(v)|} \to 0,$$

for $|H(v)| \to 0$. Now the conclusion follows since $H(v) \to 0$ as $v \to 0$ and

$$\begin{aligned} \frac{|\omega(H(u), H(v))|}{|v|} &= \frac{|\omega(H(u), H(v))|}{|H(v)|} \frac{|H(v)|}{|v|} \\ &\leq |H| \frac{|\omega(H(u), H(v))|}{|H(v)|}. \end{aligned}$$

Example 7.4 Following the ideas from Krasnoselskii [19] and using Theorem 5.1 we can prove:

Theorem 7.1 *Let $\Omega \subset \mathbf{R}^N$ be an open set, $p \in (1, \infty)$, q the conjugate exponent of p, and $F : \Omega \times \mathbf{R}^n \to \mathbf{R}$ a function with $F(x, 0) = 0$ on Ω. Assume that $F(., z)$ is measurable for all $z \in \mathbf{R}^n$ and $F(x, .)$ is continuously differentiable for a.e. $x \in \Omega$. Let $f : \Omega \times \mathbf{R}^n \to \mathbf{R}^n$ be defined as*

$$f(x, .) = \nabla F(x, .) \tag{7.6}$$

and assume f is (p, q)-Carathéodory. Then the functional $J : L^p(\Omega; \mathbf{R}^n) \to \mathbf{R}$ given by

$$J(u) = \int_\Omega F(x, u(x))\, dx$$

belongs to $C^1(L^p(\Omega; \mathbf{R}^n))$ and

$$J' = N_f,$$

that is

$$(J'(u), v) = \int_\Omega (v(x), N_f(u)(x))\, dx = \int_\Omega (v(x), f(x, u(x)))\, dx$$

for all $v \in L^p(\Omega; \mathbf{R}^n)$.

Proof. For every $u, v \in L^p(\Omega; \mathbf{R}^n)$ define the measurable function $\theta_{u,v} : \Omega \to [0, 1]$, by

$$\begin{aligned} \theta_{u,v}(x) &= \inf\{t \in [0,1] : F(x, v(x) + u(x)) \\ -F(x, v(x)) &= (u(x), f(x, v(x) + tu(x)))\}. \end{aligned}$$

Then since $F(x, 0) = 0$,

$$\begin{aligned} |J(u)| &\le \int_\Omega |F(x, u(x))|\, dx \\ &\le \int_\Omega |u(x)|\, |f(x, \theta_{u,0}(x)\, u(x))|\, dx \\ &\le |u|_p\, |N_f(\theta_{u,0} u)|_q < \infty. \end{aligned} \tag{7.7}$$

Thus J is well defined. On the other hand,

$$\begin{aligned} J(u+v)-J(u) &= \int_\Omega \left[F(x,u(x)+v(x))-F(x,u(x))\right]dx \\ &= \int_\Omega (v(x), N_f(u+\theta_{v,u}v)(x))\,dx \\ &= \int_\Omega (v(x), N_f(u)(x))\,dx \\ &\quad + \int_\Omega (v(x), N_f(u+\theta_{v,u}v)(x)-N_f(u)(x))\,dx \\ &= (N_f(u), v)+\omega(u,v). \end{aligned}$$

Here

$$\omega(u,v)=\int_\Omega (v(x), N_f(u+\theta_{v,u}v)(x)-N_f(u)(x))\,dx.$$

We have

$$|\omega(u,v)| \le |v|_p \left|N_f(u+\theta_{v,u}v)-N_f(u)\right|_q,$$

whence

$$\frac{\omega(u,v)}{|v|_p} \to 0 \quad \text{as } v \to 0,$$

since by Theorem 5.1 N_f is continuous. Thus $J'(u)=N_f(u)$. ■

Definition 7.2 A function $f:\Omega\times R^n \to R^n$ for which there exists $F:\Omega\times R^n\to R$ such that $F(x,0)=0$ a.e. on Ω, $F(.,z)$ is measurable for each $z\in R^n$, $F(x,.)$ is continuously differentiable and (7.6) holds for a.e. $x\in\Omega$, is said to be of *potential type*. In this case F is called the *potential* of f.

Remark 7.2 For $n=1$, each function $f:\Omega\times R\to R$ which satisfies the Carathéodory conditions is of potential type with

$$F(x,z)=\int_0^z f(x,y)\,dy,$$

that is, F is the primitive of f in the second variable which vanishes at zero.

Combining Examples 7.2–7.4 we obtain the following result.

Theorem 7.2 *Assume that all the assumptions of Theorem* 7.1 *hold. In addition assume that* $H : L^2(\Omega;\mathbf{R}^n) \to L^p(\Omega;\mathbf{R}^n)$ *is a bounded linear operator. Then the functional* $E : L^2(\Omega;\mathbf{R}^n) \to \mathbf{R}$, *given by*

$$E(u) = \frac{1}{2}|u|_2^2 - \int_\Omega F(x, H(u)(x))\, dx$$

is Fréchet differentiable and

$$E' = I - H^* N_f H$$

that is

$$\left(E'(u), v\right) = \left(u - H^* N_f H(u), v\right)_2 \quad \left(u, v \in L^2(\Omega;\mathbf{R}^n)\right).$$

The functional E in Theorem 7.2 is known as the *Golomb functional* generated by H and f.

7.2 Minima of Lower Semicontinuous Functionals

Now we come back to the general functional $E : U \to \mathbf{R}$, where $U \subset X$ is open. By definition $u \in U$ is a *critical point* of E if E is Fréchet differentiable at u and $E'(u) = 0$. The simplest example of critical points is that of the points of local extremum.

Proposition 7.2 *If* $u_0 \in U$ *is a point of local extremum of* E *and* E *is Fréchet differentiable at* u_0, *then* $E'(u_0) = 0$.

Proof. Assume that u_0 is a point of local minimum, i.e.,

$$E(u_0) \le E(u)$$

for all u in a neighborhood of u_0. Since E is Fréchet differentiable at u_0, we have

$$0 \le E(u_0 + v) - E(u_0) = \left(E'(u_0), v\right) + \omega(u_0, v)$$

and

$$\frac{\omega(u_0, v)}{|v|} \to 0$$

as $v \to 0$. Setting $v = tw$, where $t > 0$ and $w \in X$, $|w| = 1$, dividing by t and then letting $t \to 0^+$ we obtain

$$\left(E'(u_0), w\right) \ge 0.$$

Similarly, replacing w by $-w$ we have

$$\left(E'\left(u_0\right), -w\right) \geq 0.$$

Hence

$$\left(E'\left(u_0\right), w\right) = 0$$

for all $w \in X$, $|w| = 1$. As a result $E'\left(u_0\right) = 0$. A similar argument applies to the points of local maximum. ■

Thus a first idea in finding critical points is to look for extremal values of the functional.

Definition 7.3 A functional $E : D \subset X \to R$ is said to be *lower semicontinuous,* l.s.c. for short, on D if for every $r \in R$, the *level set*

$$(E \leq r) = \{u \in D : \, E(u) \leq r\}$$

is closed in D. E is said to be *weakly lower semicontinuous* on D if for every $r \in R$, the level set $(E \leq r)$ is closed in D with respect to the weak topology of X.

For weak topologies and additional properties of linear functional analysis we refer the reader to Brezis [5] and Yosida [44].

Proposition 7.3 *Assume* X *is a reflexive Banach space,* $D \subset X$ *is a bounded closed convex nonempty set and* $E : D \to \mathbf{R}$ *is weakly l.s.c. on* D. *Then* E *is bounded from below and attains its infimum.*

Proof. Let (u_k) be a minimizing sequence of E, i.e., $u_k \in D$ and

$$E\left(u_k\right) \to r_0 = \inf_D E \in \mathbf{R} \cup \{-\infty\}.$$

Since X is reflexive and D is bounded, we may assume, passing eventually to a subsequence, that $u_k \rightharpoonup u_0$ weakly as $k \to \infty$, where $u_0 \in X$. D being closed convex we have $u_0 \in D$. Assume $r_0 < E\left(u_0\right)$. Then there exists an $\varepsilon > 0$ and a $k_\varepsilon \in \mathbf{N}$ such that $E\left(u_k\right) \leq E\left(u_0\right) - \varepsilon$, that is

$$u_k \in \left(E \leq E\left(u_0\right) - \varepsilon\right),$$

for all $k \geq k_\varepsilon$. This, since E is weakly l.s.c. on D, implies

$$u_0 \in \left(E \leq E\left(u_0\right) - \varepsilon\right)$$

too. Thus we have obtained $E\left(u_0\right) \leq E\left(u_0\right) - \varepsilon$, a contradiction. Hence $E\left(u_0\right) = r_0$. ■

Definition 7.4 A functional $E : D \subset X \to R$ defined on an unbounded set D, is said to be *coercive* if $E(u) \to \infty$ as $|u| \to \infty$.

The analogue result of Proposition 7.3 for an unbounded set D is the following

Proposition 7.4 *Assume X is a reflexive Banach space, $D \subset X$ is an unbounded closed convex set and $E : D \to \mathbf{R}$ is coercive and weakly l.s.c. on D. Then E is bounded from below and attains its infimum.*

Proof. Fix any element u_1 in D. Since E is coercive there exists an $R > 0$ such that $E(u) > E(u_1)$ for all $u \in D$ with $|u| > R$. Consequently

$$\inf_D E = \inf_{D \cap \overline{B}_R(0)} E$$

and the conclusion follows from Proposition 7.3. ■

Theorem 7.3 *Let X be a reflexive Banach space and $E : X \to \mathbf{R}$ be coercive, weakly l.s.c. on X and Fréchet differentiable in X. Then there exists $u_0 \in X$ with*

$$E(u_0) = \inf_X E, \quad E'(u_0) = 0.$$

Proof. Apply Propositions 7.4 and 7.2. ■

The next result is a criterium of weak lower semicontinuity. It is stated in terms of convex functionals.

Definition 7.5 Let D be a convex subset of the Banach space X. A functional $E : D \to R$ is said to be *convex* if

$$E(u + t(v - u)) \leq E(u) + t(E(v) - E(u)) \tag{7.8}$$

for all $u, v \in D$, $u \neq v$ and $t \in (0, 1)$. The functional E is said to be *strictly convex* if strict inequality occurs in (7.8). The functional E is said to be *concave* if $-E$ is convex.

Proposition 7.5 *Assume X is a Banach space, $D \subset X$ is closed convex and $E : D \to \mathbf{R}$ is convex and l.s.c. on D. Then E is weakly l.s.c. on D.*

Proof. Notice each level set $(E \leq r)$ is closed convex and so is weakly closed. ■

Theorem 7.4 *Let X be a reflexive Banach space and $E : X \to \mathbf{R}$ be convex, coercive, and Fréchet differentiable in X. Then there exists $u_0 \in X$ with*

$$E(u_0) = \inf_X E, \quad E'(u_0) = 0.$$

If in addition E is strictly convex then E has a unique critical point.

Proof. Since E is Fréchet differentiable, E is continuous and so l.s.c. on X. Now the existence part follows from Proposition 7.5 and Theorem 7.3.

Assume E is strictly convex and suppose that u_1 is a critical point of E with $u_1 \neq u_0$. Then for sufficiently small $t > 0$, we have

$$\begin{aligned} E(u_0 + t(u_1 - u_0)) - E(u_0) &= t\left(E'(u_0), u_1 - u_0\right) + \omega(u_0, t(u_1 - u_0)) \\ &= \omega(u_0, t(u_1 - u_0)) \end{aligned}$$

and

$$\omega(u_0, t(u_1 - u_0)) = o(t).$$

On the other hand, since E is strictly convex, one has

$$E(u_0 + t(u_1 - u_0)) < E(u_0) + t(E(u_1) - E(u_0)) \tag{7.9}$$

for all $t \in (0,1)$. Hence

$$\omega(u_0, t(u_1 - u_0)) < t(E(u_1) - E(u_0))$$

for all $t \in (0,1)$ small enough. Dividing by t and letting $t \to 0^+$, we obtain $0 \leq E(u_1) - E(u_0)$. Similarly $0 \leq E(u_0) - E(u_1)$. Hence

$$E(u_1) = E(u_0) = \inf_X E,$$

and from (7.9) we deduce for $t \in (0,1)$,

$$E(u_0 + t(u_1 - u_0)) < \inf_X E,$$

a contradiction. ∎

7.3 Application to Hammerstein Integral Equations

We now use Theorems 7.3 and 7.4 to obtain existence and uniqueness results for the Hammerstein integral equation in $\mathbf{R}^n$

$$u(x) = \int_\Omega \kappa(x,y) f(y, u(y))\, dy \quad \text{a.e. } x \in \Omega. \tag{7.10}$$

Theorem 7.5 *Let $\Omega \subset \mathbf{R}^N$ be bounded open, $2 \le p < p_0 < \infty$, $1/p + 1/q = 1$, $1/p_0 + 1/q_0 = 1$, $\kappa : \Omega^2 \to \mathbf{R}$ and $f : \Omega \times \mathbf{R}^n \to \mathbf{R}^n$. Assume that the following conditions are satisfied:*

(i) the operator $A : L^{q_0}(\Omega; \mathbf{R}^n) \to L^{p_0}(\Omega; \mathbf{R}^n)$ given by

$$A(u)(x) = \int_\Omega \kappa(x,y)\, u(y)\, dy$$

is bounded and its restriction $A : L^2(\Omega; \mathbf{R}^n) \to L^2(\Omega; \mathbf{R}^n)$ is positive, self-adjoint and completely continuous;

(ii) f is (p,q)-Carathéodory of potential type;

(iii) the potential F of f satisfies the growth condition

$$|F(x,z)| \le \frac{a}{2}|z|^2 + b(x)|z|^{2-\gamma} + c(x) \tag{7.11}$$

for a.e. $x \in \Omega$ and all $z \in \mathbf{R}^n$, where $0 < \gamma < 2$, $b \in L^{2/\gamma}(\Omega; \mathbf{R}_+)$, $c \in L^1(\Omega; \mathbf{R}_+)$, $a \in \mathbf{R}_+$ and $a|A| < 1$ (A as operator from $L^2(\Omega; \mathbf{R}^n)$ to $L^2(\Omega; \mathbf{R}^n)$).

Then the Hammerstein equation (7.10) has at least one solution u in $L^p(\Omega; \mathbf{R}^n)$.

Proof. According to Proposition 6.6, (7.10) is equivalent to the operator equation

$$v = H^* N_f H(v), \quad v \in L^2(\Omega; \mathbf{R}^n). \tag{7.12}$$

Furthermore, this equation can be written as

$$E'(v) = 0,$$

where the functional $E : L^2(\Omega; \mathbf{R}^n) \to \mathbf{R}$ is given by

$$\begin{aligned} E(v) &= \frac{1}{2}|v|_2^2 - \int_\Omega F(x, H(v)(x))\, dx \\ &= \frac{1}{2}|v|_2^2 - JH(v). \end{aligned}$$

1) E is coercive. Indeed, by (7.11) we have

$$\begin{aligned} E(v) &= \frac{1}{2}|v|_2^2 - JH(v) \\ &\geq \frac{1}{2}|v|_2^2 - \frac{a}{2}|H(v)|_2^2 - \int_\Omega b(x)\,|H(v)(x)|^{2-\gamma}\,dx - |c|_1 . \end{aligned}$$

We have

$$\begin{aligned} |H(v)|_2^2 &= \left(A^{1/2}(v), A^{1/2}(v)\right) = (v, A(v)) \\ &\leq |A|\,|v|_2^2 \end{aligned}$$

and

$$\begin{aligned} \int_\Omega b(x)\,|H(v)(x)|^{2-\gamma}\,dx &\leq \left(\int_\Omega b(x)^{2/\gamma}\,dx\right)^{\gamma/2}\left(\int_\Omega |H(v)(x)|^2\,dx\right)^{1-\gamma/2} \\ &= b\,|H(v)|_2^{2-\gamma} \leq b\,|A|^{1-\gamma/2}\,|v|_2^{2-\gamma}, \end{aligned}$$

where $b = |b|_{2/\gamma}$. Denote $c = |c|_1$. Then

$$E(v) \geq \frac{1}{2}|v|_2^2(1 - a\,|A|) - b\,|A|^{1-\gamma/2}\,|v|_2^{2-\gamma} - c.$$

Since $1 - a\,|A| > 0$ and $2 - \gamma < 2$ we may infer that

$$E(v) \to \infty \quad \text{as} \quad |v|_2 \to \infty.$$

Thus E is coercive.

2) E is weakly l.s.c. on $L^2(\Omega; \mathbf{R}^n)$. To this end let $r > 0$ and let (v_k) be a sequence with

$$E(v_k) \leq r \ \text{ and } \ v_k \rightharpoonup v \text{ weakly as } k \to \infty.$$

Then

$$(v_k - v, u)_2 \to 0 \quad \text{as } k \to \infty$$

for every $u \in L^2(\Omega; \mathbf{R}^n)$. Since the operator A acting in $L^2(\Omega; \mathbf{R}^n)$ is completely continuous, we may assume, passing eventually to a subsequence, that $A(v_k) \to w$ in $L^2(\Omega; \mathbf{R}^n)$. We have

$$\left|A^{1/2}(v_k) - A^{1/2}(v)\right|_2^2 = (v_k - v,\ A(v_k) - A(v))_2$$

$$= (v_k - v,\ A(v_k) - w)_2 + (v_k - v,\ w - A(v))_2 \to 0$$

as $k \to \infty$, since the sequence $(v_k - v)$ is bounded, $A(v_k) - w \to 0$, and $(v_k - v, w - A(v))_2 \to 0$. It follows that

$$A^{1/2}(v_k) \to A^{1/2}(v)$$

strongly. On the other hand, H being completely continuous (see Theorem 6.1), we may assume (passing eventually to a new subsequence) that $H(v_k)$ converges in $L^p(\Omega; \mathbf{R}^n)$. Since $H(v_k) = A^{1/2}(v_k)$ the limit of $H(v_k)$ must be $H(v) = A^{1/2}(v)$. Hence

$$H(v_k) \to H(v)$$

in $L^p(\Omega; \mathbf{R}^n)$. Consequently

$$JH(v_k) \to JH(v)$$

in $\mathbf{R}$. In addition, it is well known that if $v_k \rightharpoonup v$ weakly then $|v|_2 \leq \liminf |v_k|_2$. Then since $E(v_k) \leq r$

$$\frac{1}{2}|v_k|_2^2 - r \leq JH(v_k)$$

and letting $k \to \infty$ we obtain

$$\frac{1}{2}|v|_2^2 - r \leq \liminf \frac{1}{2}|v_k|_2^2 - r \leq JH(v).$$

Hence $E(v) \leq r$. Thus, the level set $(E \leq r)$ is weakly closed. Therefore E is weakly l.s.c. on $L^2(\Omega; \mathbf{R}^n)$.

Now the conclusion follows from Theorem 7.3. ∎

If instead of the complete continuity of A as an operator acting in $L^2(\Omega; \mathbf{R}^n)$ we ask for convexity of E, by Theorem 7.4 we immediately obtain the following existence and uniqueness result.

Theorem 7.6 *Let $\Omega \subset \mathbf{R}^N$ be bounded open, $2 \leq p < p_0 < \infty$, $1/p + 1/q = 1$, $1/p_0 + 1/q_0 = 1$, $\kappa : \Omega^2 \to \mathbf{R}$ and $f : \Omega \times \mathbf{R}^n \to \mathbf{R}^n$. Assume that the following conditions are satisfied:*

(i) the operator $A : L^{q_0}(\Omega; \mathbf{R}^n) \to L^{p_0}(\Omega; \mathbf{R}^n)$ given by

$$A(u)(x) = \int_\Omega \kappa(x, y)\, u(y)\, dy$$

is bounded and its restriction $A : L^2(\Omega; \mathbf{R}^n) \to L^2(\Omega; \mathbf{R}^n)$ is positive and self-adjoint;

(ii) f is (p,q)-Carathéodory of potential type;

(iii) the potential F of f is concave in the second variable and satisfies the growth condition

$$|F(x,z)| \leq \frac{a}{2}|z|^2 + b(x)|z|^{2-\gamma} + c(x)$$

for a.e. $x \in \Omega$ and all $z \in \mathbf{R}^n$, where $0 < \gamma < 2$, $b \in L^{2/\gamma}(\Omega;\mathbf{R}_+)$, $c \in L^1(\Omega;\mathbf{R}_+)$, $a \in \mathbf{R}_+$ and $a|A| < 1$ (A as operator from $L^2(\Omega;\mathbf{R}^n)$ to $L^2(\Omega;\mathbf{R}^n)$).

Then the equation (7.10) *has a unique solution $u \in L^p(\Omega;\mathbf{R}^n)$.*

Proof. The functional $|v|_2^2/2$ is strictly convex. Also, since $F(x,.)$ is assumed concave the functional

$$-\int_\Omega F(x,H(v)(x))\,dx$$

is convex. Hence E is strictly convex and Theorem 7.4 applies. For uniqueness assume that $u_1, u_2 \in L^p(\Omega;\mathbf{R}^n)$ are solutions to (7.10), that is $u_i = AN_f(u_i)$, $i = 1,2$. Let $v_i = H^*N_f(u_i)$. Then v_i solve (7.12). Since (7.12) has a unique solution (the unique critical point of E) we have $v_1 = v_2$. Hence $H^*N_f(u_1) = H^*N_f(u_2)$. Then

$$HH^*N_f(u_1) = HH^*N_f(u_2).$$

But $HH^* = A$, so $AN_f(u_1) = AN_f(u_2)$. Thus $u_1 = u_2$. ■

We conclude this chapter by some conditions on kernel κ which guarantee (i) in Theorems 7.5:

1) $\kappa \in L^{p_0}(\Omega \times \Omega)$,
2) κ is *symmetric*, i.e., $\kappa(x,y) = \kappa(y,x)$,
3) κ is *positive semidefinite*, i.e.,

$$\int_{\Omega\times\Omega} \kappa(x,y)(u(x),u(y))\,dx\,dy \geq 0$$

for every $u \in L^2(\Omega;\mathbf{R}^n)$.

Chapter 8

The Mountain Pass Theorem and Critical Points of Saddle Type

In Chapter 9 we shall continue the investigation of the L^p solutions of the Hammerstein integral equations under the assumption that $f(x, 0) = 0$, that is, the null function is a solution. We are now interested in non-null solutions. The technique we use is based on the so called mountain pass theorem of Ambrosetti–Rabinowitz [3]. By this method one can establish the existence of a critical point u of the functional E which in general is not an extremum point of E, and has the property that in any neighborhood of u there are points v and w with $E(v) < E(u) < E(w)$. Such a critical point is said to be a *saddle point* of E.

In this chapter we first present the Ambrosetti–Rabinowitz theorem with a proof based on Ekeland's principle. Then we present Schechter's version [40] of the mountain pass theorem, which guarantees the existence of a critical point in a bounded region of the space. The proof is based on deformation arguments and uses the notion of flow associated with a locally Lipschitz map and a generalized pseudo-gradient lemma.

8.1 The Ambrosetti–Rabinowitz Theorem

Definition 8.1 A functional $E \in C^1(X)$ is said to satisfy the *Palais–Smale condition*, for short the (PS) condition, if any sequence of elements

$u_k \in X$ for which

$$E(u_k) \to \mu \in \mathbf{R}, \quad E'(u_k) \to 0 \tag{8.1}$$

as $k \to \infty$, has a convergent subsequence.

We note that the (PS) condition is a compactness property. It asks that for any sequence $(u_k)_{k \geq 1}$ satisfying (8.1), the set $\{u_k : k \geq 1\}$ is relatively compact.

Theorem 8.1 (Ambrosetti–Rabinowitz) *Let X be a Banach space and $E \in C^1(X)$. Assume that there exist $u_0, u_1 \in X$ and r with $|u_0| < r < |u_1|$ such that*

$$\max\{E(u_0), E(u_1)\} < \inf\{E(u) : u \in X, \ |u| = r\}.$$

Let

$$\Gamma = \{\gamma \in C([0,1]; X) : \gamma(0) = u_0, \ \gamma(1) = u_1\} \tag{8.2}$$

and

$$c = \inf_{\gamma \in \Gamma} \max_{t \in [0,1]} E(\gamma(t)). \tag{8.3}$$

Then there exists a sequence of elements $u_k \in X$ such that

$$E(u_k) \to c, \quad E'(u_k) \to 0 \quad as \quad k \to \infty.$$

If, in addition, E satisfies the (PS) condition then there exists an element $u \in X \setminus \{u_0, u_1\}$ with

$$E(u) = c, \quad E'(u) = 0. \tag{8.4}$$

Remark 8.1 Notice Γ is the set of all continuous paths joining u_0 and u_1. Roughly speaking, the mountain pass theorem says that if we are at the point u_0 of altitude $E(u_0)$ located in a cauldron surrounded by high mountains, and we wish to reach to a point u_1 of altitude $E(u_1)$, over there the mountains, we can find a path going from u_0 to u_1, through a mountain pass. To find a mountain pass we have to choose a path which mounts the least.

The proof that we present here relies on Ekeland's variational principle [12].

Lemma 8.1 (Ekeland) *Let (Z, d) be a complete metric space and let $\psi : Z \to \mathbf{R}$ be a lower semicontinuous function bounded from below. Then given $\varepsilon > 0$ and $u_0 \in Z$ there exists a point $u \in Z$ such that*

$$\psi(v) - \psi(u) + \varepsilon d(u, v) \geq 0 \quad \textit{for all } v \in Z, \tag{8.5}$$

$$\psi(u) \leq \psi(u_0) - \varepsilon d(u, u_0). \tag{8.6}$$

Proof. Since εd is an equivalent metric on Z, we may assume that $\varepsilon = 1$. For each $u \in Z$ we consider the set

$$Z(u) = \{v \in Z : \psi(v) - \psi(u) + d(u, v) \leq 0\}.$$

Clearly $u \in Z(u)$ and it is easily seen that if $v \in Z(u)$ then $Z(v) \subset Z(u)$. Let (ε_k) be a sequence of real numbers $\varepsilon_k > 0$ with $\varepsilon_k \to 0$ as $k \to \infty$, and let (u_k) be a sequence with

$$u_{k+1} \in Z(u_k), \quad \psi(u_{k+1}) \leq \inf_{Z(u_k)} \psi + \varepsilon_{k+1}$$

for $k = 0, 1, \ldots$. Since $u_{k+1} \in Z(u_k)$ we have

$$\psi(u_{k+1}) - \psi(u_k) + d(u_{k+1}, u_k) \leq 0. \tag{8.7}$$

Hence the sequence $(\psi(u_k))$ is decreasing. It is also bounded since ψ is bounded from below. Thus $(\psi(u_k))$ is convergent. On the other hand, from (8.7) we obtain

$$\psi(u_m) - \psi(u_k) + d(u_m, u_k) \leq 0 \quad \text{for } k < m.$$

It follows that (u_k) is a Cauchy sequence. Let $u \in Z$ be its limit. Since $u_k \in Z(u_0)$ we have

$$\psi(u_k) - \psi(u_0) + d(u_k, u_0) \leq 0,$$

and letting $k \to \infty$ we obtain (8.6). To prove (8.5) let $v \in Z$ be an arbitrary element of Z. Two cases are possible:

1) $v \in \cap Z(u_k)$. Then

$$\psi(u_{k+1}) \leq \inf_{Z(u_k)} \psi + \varepsilon_{k+1} \leq \psi(v) + \varepsilon_{k+1}$$

for all k. From this, using the lower semicontinuity of ψ we deduce $\psi(u) \leq \psi(v)$, and so (8.5).

2) $v \notin \cap Z(u_k)$. Then there exists an m such that $v \notin Z(u_k)$ for all $k \geq m$. Consequently

$$\psi(v) - \psi(u_k) + d(v, u_k) > 0, \quad k \geq m.$$

Letting $k \to \infty$ we obtain (8.5). ■

Here are two useful consequences of Ekeland's principle.

Corollary 8.1 *Under the assumptions of Lemma 8.1, for each $\varepsilon > 0$ there exists an element $u \in Z$ such that* (8.5) *holds and*

$$\psi(u) \leq \inf_Z \psi + \varepsilon. \tag{8.8}$$

Proof. Apply Lemma 8.1 to an element $u_0 \in Z$ with $\psi(u_0) \leq \inf_Z \psi + \varepsilon$. Then (8.8) immediately follows from (8.6). ■

Corollary 8.2 *Under the assumptions of Lemma 8.1, if Z is a Banach space with norm $|.|$, and ψ is a C^1 functional there exists a sequence (u_k) with*

$$\psi(u_k) \to \inf_Z \psi$$

(i.e., (u_k) is a minimizant sequence of ψ) and

$$\psi'(u_k) \to 0.$$

Proof. For $\varepsilon = 1/k$, by Corollary 8.1 there is an element $u_k \in Z$ such that

$$\psi(v) - \psi(u_k) + \frac{1}{k}|u_k - v| \geq 0, \quad v \in Z, \tag{8.9}$$

$$\psi(u_k) \leq \inf_Z \psi + \frac{1}{k}.$$

Take an arbitrary $w \in Z$ and apply (8.9) to $v = u_k + tw$, $t \in \mathbf{R}$. We obtain

$$\psi(u_k + tw) - \psi(u_k) + \frac{1}{k}|t|\,|w| \geq 0.$$

It follows that for $|t|$ small enough we have

$$t(\psi'(u_k), w) + o(|t|) + \frac{1}{k}|t|\,|w| \geq 0.$$

For $t > 0$, $t \to 0$ we deduce

$$(\psi'(u_k), w) \geq -\frac{1}{k}|w|,$$

whilst for $t < 0,\ t \to 0$ we obtain

$$\left(\psi'(u_k), w\right) \leq \frac{1}{k}|w|.$$

Hence

$$\left|\left(\psi'(u_k), w\right)\right| \leq \frac{1}{k}|w|, \quad w \in Z.$$

Therefore

$$\left|\psi'(u_k)\right| \leq \frac{1}{k}$$

whence $\psi'(u_k) \to 0$ as $k \to \infty$. ∎

If X is a Banach space and $v \in X^*$ then it is a simple consequence of the definition of the norm of v, that for each $\varepsilon > 0$ there exists $u \in X$ with

$$|u| \leq 1 \quad \text{and} \quad (v, u) > |v| - \varepsilon.$$

The next lemma guarantees that if v depends continuously on a parameter t then the corresponding element u can be choosen so that it depends continuously on t as well.

Lemma 8.2 *Let X be a Banach space and $f \in C([0,1]; X^*)$. Then there exists a function $\varphi \in C([0,1]; X)$ such that*

$$|\varphi(t)| \leq 1, \quad (f(t), \varphi(t)) > |f(t)| - \varepsilon$$

for all $t \in [0,1]$.

Proof. Let $t_0 \in [0,1]$. According to the above remark there exists $u_0 \in X$ with $|u_0| \leq 1$ and $(f(t_0), u_0) > |f(t_0)| - \varepsilon$. Let

$$U(t_0) = \{t \in [0,1] : (f(t), u_0) > |f(t)| - \varepsilon\}.$$

Clearly $t_0 \in U(t_0)$ and since f is continuous $U(t_0)$ is open in $[0,1]$. Since $[0,1] = \bigcup_{t \in [0,1]} U(t)$, there exists a finite open covering

$$\{U(t_1), U(t_2), ..., U(t_n)\}$$

of $[0,1]$. Let $u_i,\ i = 1, 2, ..., n$ be the corresponding elements, i.e.,

$$U(t_i) = \{t \in [0,1] : (f(t), u_i) > |f(t)| - \varepsilon\}.$$

If $\rho_i(t) = \operatorname{dist}(t, [0,1] \setminus U(t_i))$ define

$$\zeta_i(t) = \frac{\rho_i(t)}{\sum_{j=1}^{n} \rho_j(t)}.$$

Notice $\zeta_i : [0,1] \to [0,1]$ is continuous, $\zeta_i(t) \neq 0$ if and only if $t \in U(t_i)$, and $\sum_{i=1}^{n} \zeta_i(t) = 1$ for all $t \in [0,1]$. Now define

$$\varphi(t) = \sum_{i=1}^{n} \zeta_i(t)\, u_i.$$

Clarly $\varphi \in C([0,1]; X)$ and

$$|\varphi(t)| \leq \sum_{i=1}^{n} \zeta_i(t)\, |u_i| \leq \sum_{i=1}^{n} \zeta_i(t) = 1.$$

Also it follows easily that φ satisfies $(f(t), \varphi(t)) > |f(t)| - \varepsilon$ on $[0,1]$. ■

Proof of Theorem 8.1. We apply Corollary 8.1 to $Z = \Gamma$ given by (8.2) and to the functional $\psi : \Gamma \to \mathbf{R}$ defined by

$$\psi(\gamma) = \max_{t \in [0,1]} E(\gamma(t)).$$

This functional is lower semicontinuous and bounded from below by c (given by (8.3)). It follows that for every $\varepsilon > 0$ there exists a $\gamma_\varepsilon \in \Gamma$ with

$$\psi(\eta) - \psi(\gamma_\varepsilon) + \varepsilon d(\eta, \gamma_\varepsilon) \geq 0, \quad \eta \in \Gamma, \tag{8.10}$$

$$c \leq \psi(\gamma_\varepsilon) \leq \inf_{\Gamma} \psi + \varepsilon = c + \varepsilon.$$

Let

$$\Lambda_\varepsilon = \{t \in [0,1] : E(\gamma_\varepsilon(t)) = \psi(\gamma_\varepsilon)\}.$$

For concluding the proof it is sufficient to show that there exists a $t_\varepsilon \in \Lambda_\varepsilon$ with $|E'(\gamma_\varepsilon(t_\varepsilon))| < 2\varepsilon$. To this end we apply Lemma 8.2 to the function $f \in C([0,1]; X^*)$,

$$f(t) = E'(\gamma_\varepsilon(t)).$$

Hence there exists a function $\varphi \in C([0,1]; X)$ such that $|\varphi(t)| \leq 1$ and

$$\left(E'(\gamma_\varepsilon(t)), \varphi(t)\right) > \left|E'(\gamma_\varepsilon(t))\right| - \varepsilon$$

on $[0,1]$. In (8.10) take $\eta = \gamma_\varepsilon - \lambda w$ with $\lambda > 0$ and

$$w(t) = \zeta(t)\, \varphi(t),$$

where $\zeta : [0,1] \to [0,1]$ is continuous, $\zeta(t) = 1$ on Λ_ε, and $\zeta(0) = \zeta(1) = 0$. We have

$$d(\eta, \gamma_\varepsilon) = \lambda\, |w| \leq \lambda,$$

$$\psi(\eta) = \max_{t\in[0,1]} E(\eta(t)) = E(\eta(t_\lambda))$$

for some $t_\lambda \in [0,1]$. Hence

$$E(\eta(t_\lambda)) - \max_{t\in[0,1]} E(\gamma_\varepsilon(t)) + \varepsilon\lambda \geq 0.$$

Since

$$E(\eta(t_\lambda)) - E(\gamma_\varepsilon(t_\lambda)) = -\lambda\left(E'(\gamma_\varepsilon(t_\varepsilon)), w(t_\lambda)\right) + o(\lambda)$$

we deduce that

$$-\left(E'(\gamma_\varepsilon(t_\lambda)), w(t_\lambda)\right) + \varepsilon + \frac{1}{\lambda}o(\lambda) \geq 0.$$

We may assume that $t_\lambda \to t_\varepsilon \in \Lambda_\varepsilon$ as $\lambda \to 0$. Then

$$-\left(E'(\gamma_\varepsilon(t_\varepsilon)), \varphi(t_\varepsilon)\right) + \varepsilon \geq 0.$$

Thus

$$\left|E'(\gamma_\varepsilon(t_\varepsilon))\right| - \varepsilon < \left(E'(\gamma_\varepsilon(t_\varepsilon)), \varphi(t_\varepsilon)\right) \leq \varepsilon,$$

whence $|E'(\gamma_\varepsilon(t_\varepsilon))| < 2\varepsilon$. ■

Notice that Theorem 8.1 does not give any information about the localization of the critical point u satisfying (8.4) (of saddle type). This justifies the following question: what additional hypotheses will guarantee the existence of a critical point of saddle type in a given region of the space. Some answers to this question can be found in Pucci–Serrin [35], Guo–Sun–Qi [16], Ghoussoub–Preiss [14], Schechter [39], [40], Schechter–Tintarev [41], Ma [22], Frigon [13] and Liu–Sun [21]. In Section 8.3 we shall present Schechter's bounded mountain pass theorem in a somewhat particular case sufficient for our purposes.

8.2 Flows and Generalized Pseudo-Gradients

In this section we state and prove the lemmas that will be used in Section 8.3, in the proof of Schechter's mountain pass theorem.

Lemma 8.3 *Let X be a Banach space and $W : X \to X$ be a locally Lipschitz map such that*

$$|W(u)| \leq M, \quad u \in X.$$

Then for each $u \in X$ *there exists a unique function* $\sigma(u,.) \in C^1(\mathbf{R};X)$ *such that*

$$\begin{cases} \frac{d\sigma(u,t)}{dt} = W(\sigma(u,t)), & t \in \mathbf{R}, \\ \sigma(u,0) = u. \end{cases} \tag{8.11}$$

In addition, $\sigma \in C(X \times \mathbf{R}; X)$ *and*

$$\sigma(u, t+s) = \sigma(\sigma(u,s), t)$$

for all $u \in X$ *and* $t,\ s \in \mathbf{R}$.

Proof. Let $u \in X$ be fixed and let $r > 0$ be such that W is Lipschitz in $\overline{B}_r(u)$ with the Lipschitz constant $L > 0$. Let

$$t_1 < \frac{r}{rL + |W(u)|}$$

and

$$K = \{v \in C([0,t_1]; X) : |v(t) - u| \le r \text{ on } [0,t_1]\}.$$

Let $A : K \to K$ be given by

$$A(v)(t) = u + \int_0^t W(v(s))\,ds.$$

From

$$\begin{aligned} |A(v)(t) - u| &\le \int_0^t |W(v(s))|\,ds \\ &\le \int_0^t (|W(v(s)) - W(u)| + |W(u)|)\,ds \\ &\le (rL + |W(u)|)\,t_1 \\ &< r, \end{aligned}$$

($t \in [0,t_1]$) we have that A is well defined. In addition we can easily see that A is a contraction with Lipschitz constant $Lt_1 < 1$. By Banach's contraction principle, A has a unique fixed point $\sigma(u,.) \in C([0,t_1]; X)$. Clearly $\sigma(u,.) \in C^1([0,t_1]; X)$ and satisfies (8.11) for $t \in [0,t_1]$. Let $t_+ \in (0,\infty]$ be the supremum of all numbers t_1 such that (8.11) has a solution on $[0,t_1]$. If $t_+ < \infty$ then we take any sequence $t_k < t_+$, with $t_k \to t_+$ and we obtain

$$\begin{aligned} |\sigma(u,t_j) - \sigma(u,t_k)| &= \left| \int_{t_k}^{t_j} W(\sigma(u,s))\,ds \right| \\ &\le M\,|t_j - t_k|. \end{aligned}$$

Hence the sequence $\sigma(u, t_k)$ is Cauchy and so convergent to some w. Now again by the local existence theorem we can uniquely solve

$$\frac{d\sigma(u,t)}{dt} = W(\sigma(u,t)), \quad \sigma(u, t_+) = w$$

in a certain interval $[t_+, t_+ + \delta)$. This will contradict the definition of t_+. Thus $t_+ = \infty$. A similar argument can be used for the interval $(-\infty, 0]$. ∎

Definition 8.2 The map σ given by Lemma 8.3 is called the *flow* generated by the locally Lipschitz map W.

A source of locally Lipschitz maps are the so called *pseudo-gradients.* We now present an example of pseudo-gradient owed to Schechter [40].

Lemma 8.4 *Let X be a Hilbert space and let α, θ be real numbers with $0 < \alpha < 1 - \theta$ and $\theta \geq 0$. Then for any elements $u, v \in X \setminus \{0\}$ satisfying*

$$(u, v) \leq \theta |u| |v|$$

there exists an element $h \in X$ such that

$$(u, h) \geq \alpha |u| |h|, \quad (v, h) < 0.$$

Proof. We may assume that $|u| = |v| = 1$. We look for h in the form $h = u - \beta v$ with $\beta \geq 0$. Then

$$|h| \leq 1 + \beta, \quad (v, h) \leq \theta - \beta, \quad (u, h) \geq 1 - \beta\theta.$$

Since $0 < \alpha < 1 - \theta$, we may take a $\beta > \theta$ such that $\alpha(1 + \beta) \leq 1 - \beta\theta$; for instance, set $\beta = (1 - \alpha) / (\alpha + \theta)$. This yields the desired result. ∎

Lemma 8.5 *Let X be a Hilbert space. Let $F, G : D \subset X \to X$ be continuous, $\widehat{D} = \{u \in D : G(u) \neq 0\}$ and let $D_0 \subset \widehat{D}$ be closed. Assume that*

$$F(u) \neq 0, \quad u \in D_0,$$

and that there is a $\theta \in [0, 1)$ such that

$$(F(u), G(u)) \leq \theta |F(u)| |G(u)|, \quad u \in D_0.$$

Then for each α with $0 < \alpha < 1 - \theta$, there exists a locally Lipschitz map $H : \widehat{D} \to X$ such that

$$|H(u)| \leq 1, \quad (G(u), H(u)) \geq \alpha |G(u)|, \quad u \in \widehat{D},$$

and

$$(F(u), H(u)) < 0, \quad u \in D_0.$$

Proof. Let α' be such that $\alpha < \alpha' < 1-\theta$. Consider a map $h : \widehat{D} \to X$ with $|h(u)| = 1$ for all $u \in \widehat{D}$ and with the following properties:

$$h(u) = |G(u)|^{-1} G(u), \quad u \in \widehat{D} \setminus D_0$$

and

$$(G(u), h(u)) \geq \alpha' |G(u)|, \quad (F(u), h(u)) < 0, \quad u \in D_0.$$

Such a map exists by Lemma 8.4. Notice that one has

$$(G(u), h(u)) \geq \alpha' |G(u)|, \quad u \in \widehat{D}.$$

Since F and G are continuous, for each $u \in \widehat{D}$, there is an open neighborhood $V(u)$ of u such that

$$(G(v), h(u)) \geq \alpha |G(v)|, \quad (F(v), h(u)) < 0, \quad u \in D_0$$

for all $v \in V(u)$. For $u \in \widehat{D} \setminus D_0$, take $V(u)$ so small that $V(u) \cap D_0 = \emptyset$ (recall D_0 is a closed set) and

$$(G(v), h(u)) \geq \alpha |G(v)|, \quad v \in V(u).$$

The collection $\{V(u) : u \in \widehat{D}\}$ is an open covering of $\widehat{D}$. Since $\widehat{D}$ is paracompact, this covering has a locally finite refinement $\{V_\tau\}$. Let $\{\psi_\tau\}$ be a locally Lipschitz partition of unity subordinate to this refinement, and for each τ, let $u_\tau \in \widehat{D}$ be an element for which $V_\tau \subset V(u_\tau)$. Now let $H : \widehat{D} \to X$ be given by

$$H(v) = \sum_\tau \psi_\tau(v) h(u_\tau).$$

Clearly H is locally Lipschitz. Also, for every $v \in \widehat{D}$ we have $|H(v)| \leq 1$ and

$$\begin{aligned} (G(v), H(v)) &= \sum_\tau \psi_\tau(v) (G(v), h(u_\tau)) \\ &\geq \alpha \sum_\tau \psi_\tau(v) |G(v)| \\ &= \alpha |G(v)|, \end{aligned}$$

whilst for $v \in D_0$, one has

$$\begin{aligned} (F(v), H(v)) &= \sum_\tau \psi_\tau(v) (F(v), h(u_\tau)) \\ &< 0, \end{aligned}$$

which concludes the proof. ■

8.3 Schechter's Bounded Mountain Pass Theorem

Throughout this section X will be a Hilbert space and for any $R \in (0, \infty]$, B_R will be the open set $\{u \in X : |u| < R\}$. Clearly, for $R = \infty$ one has $B_R = \overline{B}_R = X$ and $\partial B_R = \emptyset$.

Definition 8.3 Let $R \in (0, \infty]$ and $E \in C^1 \left(\overline{B}_R\right)$. For $R < \infty$, we say that E satisfies *Schechter's Palais–Smale condition*, (PS)$_R$ condition for short, if any sequence of elements $u_k \in \overline{B}_R \setminus \{0\}$ for which

$$E(u_k) \to \mu \in \mathbf{R},\ E'(u_k) - \frac{(E'(u_k), u_k)}{|u_k|^2} u_k \to 0,\ (E'(u_k), u_k) \to \nu \leq 0 \tag{8.12}$$

as $k \to \infty$, has a convergent subsequence. For $R = \infty$ we let (PS)$_\infty$ =(PS).

Remark 8.2 If E satisfies the (PS)$_R$ condition, then E satisfies the Palais–Smale condition on $\overline{B}_R$, i.e., any sequence of elements $u_k \in \overline{B}_R$ satisfying (8.1) has a convergent subsequence. For $R = \infty$ this is true by definition. Assume $R < \infty$. Then since (u_k) is bounded and $E'(u_k) \to 0$ we have $(E'(u_k), u_k) \to 0$, so $\nu = 0$. Also,

$$\left| \frac{(E'(u_k), u_k)}{|u_k|^2} u_k \right| \leq |E'(u_k)| \to 0.$$

Consequently

$$E'(u_k) - \frac{(E'(u_k), u_k)}{|u_k|^2} u_k \to 0.$$

Thus (u_k) satisfies (8.12), and so by the (PS)$_R$ condition it has a convergent subsequence.

Theorem 8.2 (Schechter) *Let X be a Hilbert space, $R \in (0, \infty]$, and $E \in C^1 \left(\overline{B}_R\right)$. Assume that for some $\nu_0 > 0$,*

$$(E'(u), u) \geq -\nu_0, \quad u \in \partial B_R \tag{8.13}$$

and that there are $u_0,\ u_1 \in \overline{B}_R$ and r with $|u_0| < r < |u_1|$ such that

$$\max\{E(u_0), E(u_1)\} < \inf\left\{E(u) : u \in \overline{B}_R,\ |u| = r\right\}. \tag{8.14}$$

Let

$$\Gamma_R = \left\{\gamma \in C\left([0,1]; \overline{B}_R\right) : \gamma(0) = u_0,\ \gamma(1) = u_1\right\}$$

and

$$c_R = \inf_{\gamma \in \Gamma_R} \max_{t \in [0,1]} E(\gamma(t)).$$

Then either there is a sequence of elements $u_k \in \overline{B}_R$ *with*

$$E(u_k) \to c_R, \quad E'(u_k) \to 0, \tag{8.15}$$

or there is a sequence of elements $u_k \in \partial B_R$ *such that*

$$E(u_k) \to c_R, \; E'(u_k) - \frac{(E'(u_k), u_k)}{R^2} u_k \to 0, \; (E'(u_k), u_k) \leq 0. \tag{8.16}$$

If in addition E *satisfies the* $(PS)_R$ *condition and*

$$E'(u) + \mu u \neq 0, \quad u \in \partial B_R, \; \mu > 0, \tag{8.17}$$

then there exists an element $u \in \overline{B}_R \setminus \{u_0, u_1\}$ *with*

$$E(u) = c_R, \quad E'(u) = 0.$$

Remark 8.3 For $R = \infty$, $\Gamma_R = \Gamma$, $c_R = c$, conditions (8.13) and (8.17) are trivial since $\partial B_R = \emptyset$, and Theorem 8.2 reduces to Theorem 8.1.

Remark 8.4 For $R < \infty$, (8.17) is the Leray–Schauder boundary condition for the operator $I - E'$, i.e., it is equivalent to

$$u \neq \lambda (I - E')(u), \quad u \in \partial B_R, \; \lambda \in (0,1).$$

The proof of Theorem 8.2 is based on the following lemma [40].

Lemma 8.6 *Assume all the assumptions of Theorem 8.2 hold. In addition assume that there are constants* θ, δ *with* $0 \leq \theta < 1$, $\delta > 0$ *such that*

$$(E'(u), u) + \theta R |E'(u)| \geq 0 \tag{8.18}$$

for all $u \in \partial B_R$ *satisfying*

$$|E(u) - c_R| \leq \delta. \tag{8.19}$$

Then there exists a sequence of elements $u_k \in \overline{B}_R$ *such that*

$$E(u_k) \to c_R, \quad E'(u_k) \to 0. \tag{8.20}$$

Proof. Assume there are no sequences satisfying (8.20). Then there would be constants δ, $m > 0$ such that

$$\left|E'(u)\right| \geq m \tag{8.21}$$

for all u in

$$Q = \left\{u \in \overline{B}_R : |E(u) - c_R| \leq 3\delta\right\}.$$

Clearly, we may assume $3\delta < c_R - \max\{E(u_0), E(u_1)\}$ and that (8.18) holds in $\widetilde{Q} = Q \cap \partial B_R$. Denote

$$\begin{aligned}
Q_0 &= \left\{u \in \overline{B}_R : |E(u) - c_R| \leq 2\delta\right\}, \\
Q_1 &= \left\{u \in \overline{B}_R : |E(u) - c_R| \leq \delta\right\}, \\
Q_2 &= \overline{B}_R \setminus Q_0, \\
\eta(u) &= \frac{d(u, Q_2)}{d(u, Q_1) + d(u, Q_2)}.
\end{aligned}$$

We have

$$\eta(u) = 1 \text{ in } \overline{Q}_1, \quad \eta(u) = 0 \text{ in } \overline{Q}_2, \quad 0 < \eta(u) < 1 \text{ otherwise.}$$

We now apply Lemma 8.5 to $F(u) = -u$, $G(u) = E'(u)$, $D = \overline{B}_R$ and $D_0 = \widetilde{Q}$. It follows that for each positive $\alpha < 1 - \theta$, there exists a locally Lipschitz map $H : \widehat{D} \to X$ (here $\widehat{D}$ means the set $\{u \in \overline{B}_R : E'(u) \neq 0\}$) such that

$$|H(u)| \leq 1, \quad \alpha\left|E'(u)\right| \leq \left(E'(u), H(u)\right), \quad u \in \widehat{D} \tag{8.22}$$

and

$$(u, H(u)) > 0, \quad u \in \widetilde{Q}. \tag{8.23}$$

Define $W : \overline{B}_R \to X$ by

$$W(u) = \begin{cases} -\eta(u) H(u) & \text{for } u \in \widehat{D} \\ 0 & \text{for } u \in \overline{B}_R \setminus \widehat{D}. \end{cases} \tag{8.24}$$

This map is locally Lipschitz and can be extended to a locally Lipschitz map on the whole of X, by setting

$$W(u) = W\left(\frac{R}{|u|}u\right) \quad \text{for } |u| > R.$$

Let σ be the flow generated by W as shows Lemma 8.3. Note $\sigma(u,.)$ does not exit $\overline{B}_R$ for every $u \in \overline{B}_R$. Indeed, we have

$$\frac{d\,|\sigma(u,t)|^2}{dt} = 2\left(\frac{d\sigma(u,t)}{dt}, \sigma(u,t)\right) = 2\left(W(\sigma), \sigma\right). \tag{8.25}$$

Assume $|\sigma(u,t)| \leq R$ for all $t \in [0,t_0)$ and $u_1 = \sigma(u,t_0) \in \partial B_R$, for some $t_0 \in \mathbf{R}_+$. If $u_1 \in \widetilde{Q}$ then by (8.23) and (8.24) we see that

$$(W(\sigma(u,t)), \sigma(u,t)) < 0$$

for t in a neighborhood of t_0. If $u_1 \notin \widetilde{Q}$ then

$$\eta(\sigma(u,t)) = 0$$

for t in a neighborhood of t_0. Hence the right hand side of (8.25) is non-positive in a neighborhood of t_0. Thus $|\sigma(u,t)|$ is nonincreasing on some interval $[t_0, t_0+\varepsilon)$. Therefore $\sigma(u,.)$ does not exit $\overline{B}_R$ for $t \in \mathbf{R}_+$.

Let us denote by E_λ the level set $(E \leq \lambda)$, i.e.,

$$E_\lambda = \left\{u \in \overline{B}_R : E(u) \leq \lambda\right\}.$$

Now we look at the composite map $E(\sigma(u,.))$. By (8.21) and (8.22) we have

$$\begin{aligned}
\frac{dE(\sigma(u,t))}{dt} &= \left(E'(\sigma(u,t)), \frac{d\sigma(u,t)}{dt}\right) \\
&= -\eta(\sigma(u,t))\left(E'(\sigma(u,t)), H(\sigma(u,t))\right) \\
&\leq -\eta(\sigma(u,t))\,\alpha\, m.
\end{aligned} \tag{8.26}$$

Let $t_1 > 2\delta/(\alpha\, m)$, and let u be any element of $E_{c_R+\delta}$. If there is a $t_0 \in [0,t_1]$ with $\sigma(u,t_0) \notin Q_1$, then

$$\begin{aligned}
E(\sigma(u,t_1)) &\leq E(\sigma(u,t_0)) \\
&< c_R - \delta.
\end{aligned}$$

Hence $\sigma(u,t_1) \in E_{c_R-\delta}$. Otherwise, $\sigma(u,t) \in Q_1$ for all $t \in [0,t_1]$, and so $\eta(\sigma(u,t)) \equiv 1$. Then (8.26) implies

$$\begin{aligned}
E(\sigma(u,t_1)) &\leq E(u) - \alpha\, m t_1 \\
&< c_R + \delta - 2\delta = c_R - \delta.
\end{aligned}$$

Thus

$$\sigma(E_{c_R+\delta}, t_1) \subset E_{c_R-\delta}. \tag{8.27}$$

Now by the definition of c_R, there is a $\gamma \in \Gamma_R$ with

$$\gamma(t) \in E_{c_R+\delta}, \quad t \in [0,1]. \tag{8.28}$$

We define a new path γ_1 joining u_0 and u_1 by

$$\gamma_1(t) = \sigma(\gamma(t), t_1), \quad t \in [0,1].$$

Since η vanishes in the neighborhood of u_0 and u_1, we have $\sigma(u_0, t) \equiv u_0$ and $\sigma(u_1, t) \equiv u_1$. Hence $\gamma_1(0) = u_0$, $\gamma_1(1) = u_1$ and so $\gamma_1 \in \Gamma_R$. On the other hand, from (8.27) and (8.28), we have

$$E(\gamma_1(t)) \leq c_R - \delta, \quad t \in [0,1],$$

which contradicts the definition of c_R. This finishes the proof of Lemma 8.6. ■

Proof of Theorem 8.2. Assume a sequence satisfying (8.16) does not exist. Then there are constants m, $\delta > 0$ such that

$$\left|E'(u) - R^{-2}(E'(u), u)\, u\right| \geq m$$

whenever

$$|E(u) - c_R| \leq \delta, \quad (E'(u), u) \leq 0, \quad u \in \partial B_R. \tag{8.29}$$

Let $\theta > 0$ be such that

$$0 < \theta^{-2} - 1 < m^2 R^2 \nu_0^{-2}.$$

Then if u satisfies (8.29) we have

$$(E'(u), u)^2 (\theta^{-2} - 1) \leq R^2 \left|E'(u) - R^{-2}(E'(u), u)\, u\right|^2 (E'(u), u)^2 \nu_0^{-2}.$$

Since $(E'(u), u)^2 \nu_0^{-2} \leq 1$ this yields

$$\begin{aligned} \theta^{-2}(E'(u), u)^2 &\leq (E'(u), u)^2 + R^2 \left|E'(u) - R^{-2}(E'(u), u)\, u\right|^2 \\ &= R^2 \left|E'(u)\right|^2. \end{aligned}$$

Hence u satisfies (8.18). Thus (8.18) holds for all u satisfying (8.29). It also holds trivially for all u with $(E'(u), u) > 0$. Hence (8.18) holds for all $u \in \partial B_R$ satisfying (8.19). Therefore Lemma 8.6 applies and guarantees the existence of a sequence (u_k) satisfying (8.20).

Finally, assume that E satisfies the $(\mathrm{PS})_R$ condition. Then we may assume, passing if necessary to a subsequence, that $u_k \to u$ for some $u \in \overline{B}_R$. Then

$$E(u) = c_R, \quad E'(u) = 0, \tag{8.30}$$

or

$$|u| = R,\ E(u) = c_R,\ E'(u) - R^{-2}(E'(u), u)\, u = 0,\ (E'(u), u) \leq 0, \tag{8.31}$$

respectively. If in (8.31) $(E'(u), u) = 0$ then u also satisfies (8.30). Assume (8.31) holds with $(E'(u), u) < 0$ and let $\mu = -R^{-2}(E'(u), u)$. Then $\mu > 0$ and $E'(u) + \mu u = 0$, which contradicts (8.17).

Notice that $u \neq u_0$ and $u \neq u_1$ since by (8.14),

$$E(u) = c_R \geq \inf\{E(w) : |w| = r\} > \max\{E(u_0), E(u_1)\}.$$

The proof is complete. ∎

Remark 8.5 Theorem 8.2 remains true if in (8.14) one has equality. Moreover, if

$$\inf\left\{E(w) : w \in \overline{B}_R,\ |w| = r\right\} = c_R$$

then a sequence (u_k) like that in Theorem 8.2 can be found such that

$$d(u_k, \partial B_r) \to 0.$$

Consequently we obtain a critical point u with $|u| = r$ (see Schechter [40], pp 103–104).

Remark 8.6 For $R = \infty$, the proof of Theorem 8.2 represents another proof of Theorem 8.1 based on deformation arguments.

The next result is also owed to Schechter (see [40], Corollary 5.4.4) and guarantees the existence of at least two distinct critical points.

Theorem 8.3 (Schechter) *Assume all the assumptions of Theorem 8.2 hold and that E satisfies the $(PS)_R$ condition. In addition assume*

$$m_R = \inf_{u \in \overline{B}_R} E(u) > -\infty.$$

Then E has at least two different critical points u, u_m such that

$$u \in \overline{B}_R \setminus \{u_0, u_1\}, \quad E(u) = c_R$$

and

$$u_m \in \overline{B}_R, \quad E(u_m) = m_R.$$

Moreover, if $E(u_1) \leq E(u_0)$ then we may assume that $u_m \neq u_0$.

For much more information on critical point theory we refer the reader to Ambrosetti [2], Benci [4], Brezis–Nirenberg [6], Kavian [18], Mawhin–Willem [23], Rabinowitz [36] and Struwe [42]. For nonsmooth critical point theory, see Canino–Degiovanni [7], Chang [8], Corvellec [10], Frigon [13] and Motreanu–Varga [27].

Chapter 9

Nontrivial Solutions of Abstract Hammerstein Equations

This chapter deals with nontrivial solvability in balls of abstract Hammerstein equations and Hammerstein integral equations in $\mathbf{R}^n$ by a variational approach. The variational method for treating Hammerstein integral equations goes back to the papers of Hammerstein [17] and Golomb [15]. For further contributions see the monographs by Krasnoselskii [19], Krasnoselskii–Zabreiko–Pustylnik–Sobolevskii [20] and Vainberg [43]. For more recent results see the papers by Moroz–Vignoli–Zabreiko [24] and Moroz–Zabreiko [25], [26]. The results presented in this chapter were adapted from Precup [31]–[34].

9.1 Nontrivial Solvability of Abstract Hammerstein Equations

Here we discuss the *abstract Hammerstein equation*

$$u = AN(u), \quad u \in Y, \tag{9.1}$$

where Y is a Banach space, $N : Y \to Y^*$ and $A : Y^* \to Y$ is linear. Assume that A splits into

$$\begin{cases} A = HH^* \text{ with } H : X \to Y \text{ and } H^* : Y^* \to X, \\ \text{where } X \text{ is a Hilbert space.} \end{cases} \tag{9.2}$$

Then (9.1) can be converted into an equation in X, namely

$$v = H^* N H (v), \quad v \in X. \tag{9.3}$$

Indeed, if u solves (9.1) then $v = H^* N (u)$ is a solution of (9.3), and conversely if v solves (9.3) then $u = H (v)$ is a solution of (9.1). Moreover, H realizes an one-to-one correspondence between the solution sets of the two equations (use a similar argument as in the proof of Proposition 6.6). If, in addition, we assume

$$N = J' \text{ for some } J \in C^1 (Y; \mathbf{R}), \ J(0) = 0, \tag{9.4}$$

and

$$H \text{ is bounded linear and } H^* \text{ is the adjoint of } H, \tag{9.5}$$

then (9.3) is equivalent to the critical point problem

$$E'(v) = 0, \quad v \in X$$

for the *energy (or Golomb) functional*

$$E : X \to \mathbf{R}, \quad E(v) = \frac{1}{2} |v|_X^2 - JH(v).$$

Here $|.|_X$ stands for the norm of X. Notice $E \in C^1 (X; \mathbf{R})$ and

$$E'(v) = v - H^* N H (v), \quad v \in X.$$

We now state an existence principle for (9.1) in a ball of X.

Theorem 9.1 *Assume* (9.2), (9.4) *and* (9.5). *Assume that* $N(0) = 0$ *and the functional* $(N(.), .)$ *sends bounded sets into upper bounded sets. In addition assume that there are* $v_1 \in X \setminus \{0\}$, $r \in (0, |v_1|)$ *and* $R \geq |v_1|$ *such that the following conditions are satisfied:*

$$\max \left\{ 0, \frac{1}{2} |v_1|_X^2 - JH(v_1) \right\} < \inf \left\{ \frac{1}{2} |v|_X^2 - JH(v) : |v|_X = r \right\}, \tag{9.6}$$

$$v \neq \lambda H^* N H (v) \quad \textit{for } |v|_X = R, \ \lambda \in (0,1), \tag{9.7}$$

$$E \textit{ satisfies the } (PS)_R \textit{ condition.} \tag{9.8}$$

Then there exists a $v \in X \setminus \{0\}$ *with* $|v|_X \leq R$ *such that* $u = H(v)$ *is a non-zero solution of* (9.1).

Proof. We apply Theorem 8.2. It is clear that (9.6) guarantees (8.14), where $u_0 = 0$, $u_1 = v_1$, and (9.7) is equivalent to (8.17). Also (8.13) holds. Indeed, for $v \in \partial B_R$ we have

$$\begin{aligned} (E'(v), v) &= (v - H^* NH(v), v) \\ &= R^2 - (NH(v), H(v)), \end{aligned}$$

and since H is bounded and $(N(.), .)$ sends bounded sets into upper bounded sets, there exists a $\nu_0 > 0$ such that

$$(NH(v), H(v)) \leq R^2 + \nu_0, \quad v \in \partial B_R,$$

equivalently, (8.13) holds. Thus, also taking into account (9.8), all the assumptions of Theorem 8.2 are satisfied. ■

From Theorem 8.3 we can derive the existence of two solutions for (9.1).

Theorem 9.2 *Assume all the assumptions of Theorem 9.1 hold. In addition assume that J sends bounded sets into upper bounded sets and*

$$\frac{1}{2} |v_1|_X^2 - JH(v_1) \leq 0.$$

Then (9.1) *has at least two non-zero solutions $u \in Y$ of the form $u = H(v)$ with $|v|_X \leq R$.*

Proof. Apply Theorem 8.3. ■

Remark 9.1 (The energy space) Assume (9.2), (9.4), (9.5) hold and $H : X \to Y$ is injective. The space

$$X_H := H(X) \subset Y$$

is a Hilbert space under the scalar product $(.,.)_H$ defined by

$$(H(v_1), H(v_2))_H = (v_1, v_2) \quad (v_1, v_2 \in X),$$

and is called the *energy space* generated by H. The corresponding *energy norm* is given by

$$|H(v)|_H = |v| \quad (v \in X).$$

Since $H : X \to Y$ is bounded linear,

$$|H(v)|_Y \leq |H| \, |v|_X = |H| \, |H(v)|_H$$

for all $v \in X$. Hence

$$\text{the embedding } \ X_H \subset Y \ \text{ is continuous.} \tag{9.9}$$

If $u \in Y$ is a solution of (9.1), then $u = H(v)$ for $v = H^* N(u) \in X$, so $u \in X_H$. Thus the solutions of (9.1) belong to the energy space X_H. In addition, $u \in X_H$ solves (9.1) if and only if $u = H(v)$ and $v = H^* N(u)$ solves (9.3) in X, that is

$$(v, \xi) = (H^* N H(v), \xi), \quad \xi \in X,$$

or, equivalently,

$$(H(v), H(\xi))_H = (N H(v), H(\xi)), \quad \xi \in X.$$

Hence $u \in X_H$ is a solution of (9.1) if and only if u satisfies the variational identity

$$(u, w)_H = (N(u), w), \quad w \in X_H.$$

Now define the functional $E : X_H \to R$, by

$$E(u) = \frac{1}{2} |u|_H^2 - J(u), \quad u \in X_H.$$

Since $J \in C^1(Y; \mathbf{R})$, by (9.9), $J \in C^1(X_H; \mathbf{R})$ and has the same derivative N. Hence

$$(E'(u), w) = (u, w)_H - (N(u), w) \quad (u, w \in X_H).$$

Thus $u \in X_H$ is a solution of (9.1) if and only if u is a critical point of E.

Therefore, a direct variational approach to (9.1) is possible in the energy space X_H.

9.2 Nontrivial Solutions of Hammerstein Integral Equations

In this section we shall specialize Theorems 9.1 and 9.2 for the Hammerstein integral equation in $\mathbf{R}^n$

$$u(x) = \int_\Omega \kappa(x, y) f(y, u(y)) \, dy \quad \text{a.e. on } \Omega. \tag{9.10}$$

Theorem 9.3 *Let $\Omega \subset \mathbf{R}^N$ be bounded open, $2 \leq p < p_0 < \infty$, $1/p+1/q = 1$, $1/p_0 + 1/q_0 = 1$, $\kappa : \Omega^2 \to \mathbf{R}$ and $f : \Omega \times \mathbf{R}^n \to \mathbf{R}^n$. Assume that the following conditions are satisfied:*

(i) the operator $A : L^{q_0}(\Omega; \mathbf{R}^n) \to L^{p_0}(\Omega; \mathbf{R}^n)$ given by

$$A(u)(x) = \int_\Omega \kappa(x,y)\, u(y)\, dy$$

is bounded and its restriction $A : L^2(\Omega; \mathbf{R}^n) \to L^2(\Omega; \mathbf{R}^n)$ is positive, self-adjoint, and completely continuous;

(ii) f is (p,q)-Carathéodory of potential F and $f(x,0) = 0$ a.e. on Ω;

(iii) there are $v_1 \in L^2(\Omega; \mathbf{R}^n) \setminus \{0\}$ and $r \in (0, |v_1|_2)$ such that

$$\max\left\{0, \frac{1}{2}|v_1|_2^2 - \int_\Omega F(y, H(v_1)(y))\, dy\right\} \tag{9.11}$$

$$< \inf\left\{\frac{1}{2}|v|_2^2 - \int_\Omega F(y, H(v)(y))\, dy : v \in L^2(\Omega; \mathbf{R}^n),\ |v|_2 = r\right\};$$

(iv) there is $R \geq |v_1|_2$ such that

$$H(v) \neq \lambda \int_\Omega \kappa(., y)\, f(y, H(v)(y))\, dy \tag{9.12}$$

for every $v \in L^2(\Omega; \mathbf{R}^n)$ with $|v|_2 = R$ and all $\lambda \in (0,1)$.

Then the Hammerstein equation (9.10) *has at least one non-zero solution u in $L^p(\Omega; \mathbf{R}^n)$ of the form $u = H(v)$ with $v \in L^2(\Omega; \mathbf{R}^n)$, $|v|_2 \leq R$.*

Proof. Apply Theorem 9.1 to $N = N_f$ and J given by

$$J(u) = \int_\Omega F(y, u(y))\, dy \quad (u \in L^p(\Omega; \mathbf{R}^n)).$$

Since N_f is a bounded operator, the map $(N_f(.), .)$ sends bounded sets into bounded sets. By (iii), (iv), conditions (9.6) and (9.7) hold trivially. It remains to show that the attached functional E satisfies the $(\mathrm{PS})_R$ condition. To do this let (v_k) be any sequence of elements in $L^2(\Omega; \mathbf{R}^n)$ with

$$0 < |v_k|_2 \leq R$$

satisfying

$$E(v_k) \to \mu \in \mathbf{R}, \quad (E'(v_k), v_k)_2 \to \nu \leq 0$$

and

$$E'(v_k) - \frac{(E'(v_k), v_k)_2}{|v_k|_2^2} v_k \to 0. \tag{9.13}$$

We may assume that $|v_k|_2 \to a$, for some $a \in [0, R]$. If $a = 0$ we have finished. Assume $a \in (0, R]$. Then

$$\frac{(E'(v_k), v_k)_2}{|v_k|_2^2} \to \frac{\nu}{a^2} \in (-\infty, 0].$$

On the other hand, (v_k) being bounded and H being completely continuous (see Theorem 6.1), we may assume, passing eventually to a subsequence, that $(H(v_k))$ converges. Then since N_f and H^* are continuous we deduce that $(H^* N_f H(v_k))$ converges too. Now from (9.13) which can be written as

$$v_k - H^* N_f H(v_k) - \frac{(E'(v_k), v_k)_2}{|v_k|_2^2} v_k \to 0$$

we infer that the corresponding subsequence of (v_k) is convergent. Thus E satisfies the $(\mathrm{PS})_R$ condition. ■

Theorem 9.2 yields the following result.

Theorem 9.4 *Assume all the assumptions of Theorem 9.3 hold. In addition assume that*

$$\frac{1}{2}|v_1|_2^2 - \int_\Omega F(y, H(v_1)(y))\, dy \le 0. \tag{9.14}$$

Then (9.1) *has at least two non-zero solutions* $u \in L^p(\Omega; \mathbf{R}^n)$ *of the form* $u = H(v)$ *with* $|v|_2 \le R$.

Proof. Notice J maps bounded sets into bounded sets as follows from (7.7). The result is now a simple consequence of Theorem 9.2. ■

Remark 9.2 If $p > 2$, a sufficient condition for (iii) is the following one:

(iii*) there is a $v_1 \in L^2(\Omega; \mathbf{R}^n) \setminus \{0\}$ such that (9.14) holds,

$$|f(x, z)| \le a\left(1 + |z|^{p-1}\right) \tag{9.15}$$

for a.e. $x \in \Omega$ and all $z \in R^n$, where $a \in R_+$, and

$$\lim_{z \to 0} \frac{|f(x, z)|}{|z|} = 0 \tag{9.16}$$

uniformly for a.e. $x \in \Omega$.

Indeed, (9.16) implies that for any given $\varepsilon > 0$ there is a $\delta = \delta(\varepsilon) > 0$ such that

$$|f(x,z)| \leq \varepsilon |z| \quad \text{for a.e. } x \in \Omega \text{ and } |z| < \delta.$$

For $|z| \geq \delta$, from (9.15) we deduce

$$\begin{aligned} |f(x,z)| &\leq a\left(\frac{|z|}{\delta} + |z|^{p-1}\right) \\ &\leq a\left(\left(\frac{|z|}{\delta}\right)^{p-1} + |z|^{p-1}\right) \\ &= c(\varepsilon)\,|z|^{p-1}. \end{aligned}$$

Here $c(\varepsilon) = a\left(\delta^{1-p} + 1\right) \geq 0$. Hence

$$|f(x,z)| \leq \varepsilon |z| + c(\varepsilon)\,|z|^{p-1}$$

for a.e. $x \in \Omega$ and all $z \in R^n$. Now using (7.7) we obtain

$$\begin{aligned} E(v) &= \frac{1}{2}|v|_2^2 - JH(v) \\ &\geq \frac{1}{2}|v|_2^2 - \int_\Omega |H(v)(x)|\,\left|f\left(x, \theta_{H(v),0}(x)\,H(v)(x)\right)\right| dx \\ &\geq \frac{1}{2}|v|_2^2 - \int_\Omega |H(v)|\left(\varepsilon\,|H(v)| + c(\varepsilon)\,|H(v)|^{p-1}\right) dx \\ &= \frac{1}{2}|v|_2^2 - \varepsilon\,|H(v)|_2^2 - c(\varepsilon)\,|H(v)|_p^p. \end{aligned}$$

Since

$$\begin{aligned} |H(v)|_2^2 &= (H(v), H(v))_2 = (H^*H(v), v)_2 \\ &\leq |H^*|\,|H|\,|v|_2^2 \end{aligned}$$

and

$$|H(v)|_p \leq c\,|v|_2$$

because H is bounded, we deduce

$$\begin{aligned} E(v) &\geq \frac{1}{2}|v|_2^2 - \varepsilon\,|H^*|\,|H|\,|v|_2^2 - \overline{c}(\varepsilon)\,|v|_2^p \\ &= \left(\frac{1}{2} - \varepsilon\,|H^*|\,|H| - \overline{c}(\varepsilon)\,|v|_2^{p-2}\right)|v|_2^2. \end{aligned}$$

Now we choose any $\varepsilon > 0$ such that

$$\frac{1}{2} - \varepsilon |H^*| |H| > 0$$

and an $r > 0$ small enough so that

$$r < |v_1|_2, \quad \frac{1}{2} - \varepsilon |H^*| |H| - \overline{c}(\varepsilon) r^{p-2} > 0$$

(recall $p > 2$). Then

$$E(v) > 0 \quad \text{for} \quad |v|_2 = r.$$

This together with (9.14) guarantees (9.11).

Remark 9.3 Assume $H : L^2(\Omega; \mathbf{R}^n) \to L^p(\Omega; \mathbf{R}^n)$ is bijective. Then, by the theorem on the continuity of the inverse operator (see Brezis [5], Corollary II.6) there are constants $\alpha, \beta > 0$ such that

$$\alpha |v|_2 \leq |H(v)|_p \leq \beta |v|_2, \quad v \in L^2(\Omega; \mathbf{R}^n).$$

Now a sufficient condition for *(iv)* is to exist an $R \geq |v_1|_2$ with

$$|A|_{q,p} \left(|g|_q + c\beta^{p-1} R^{p-1} \right) \leq \alpha R.$$

Here $|A|_{q,p}$ is norm of A as operator from $L^q(\Omega; \mathbf{R}^n)$ to $L^p(\Omega; \mathbf{R}^n)$, whilst $g \in L^q(\Omega; \mathbf{R}_+)$ and $c \in R_+$ come from the (p,q)-Carathéodory property of f, i.e.,

$$|f(x,z)| \leq g(x) + c|z|^{p-1}$$

for a.e. $x \in \Omega$ and all $z \in R^n$.

Indeed, if $v \in L^2(\Omega; \mathbf{R}^n)$ satisfies $|v|_2 = R$ then

$$\begin{aligned} |AN_f H(v)|_p &\leq |A|_{q,p} |N_f H(v)|_q \\ &\leq |A|_{q,p} \left(|g|_q + c |H(v)|_p^{p-1} \right) \\ &\leq |A|_{q,p} \left(|g|_q + c\beta^{p-1} R^{p-1} \right) \\ &\leq \alpha R \\ &\leq |H(v)|_p. \end{aligned}$$

This guarantees (9.12) for all $\lambda \in (0,1)$.

Similar results for the existence of solutions in a ball of $L^p(\Omega; \mathbf{R}^n)$ are presented in Precup [31], [32] by means of a mountain pass theorem on closed convex sets owed to Guo–Sun–Qi [16].

9.3 A Localization Result for Nontrivial Solutions

The results presented in the previous two sections can also be viewed as theorems of localization of the nontrivial solutions in balls. In this section we try to localize solutions in a shell. The main idea for the localization of a solution to equation (9.1) in a shell of X consists in finding two numbers R_0 and R with $0 < R_0 < R$ such that all the assumptions of Theorem 9.1 hold in the ball of X of radious R and (9.3) has no solutions v satisfying $0 < |v|_X < R_0$. Then, of course, (9.1) will have a solution $u \in Y$ of the form $u = H(v)$, with $v \in X$ and

$$R_0 \leq |v|_X \leq R.$$

Here we present just an example of a problem for which such a localization is possible. The ideas in this section were adapted from Precup [34]. Naturally, an open problem is to give extensions for other nonlinear problems.

The problem we deal with is the semi-linear *Dirichlet problem*

$$\begin{cases} -\Delta u = |u|^{p-2}\, u & \text{in } \Omega, \\ u = 0 & \text{on } \partial\Omega. \end{cases} \tag{9.17}$$

It is known (see Struwe [42], Theorem II.6.6) that if $\Omega \subset \mathbf{R}^N$ is a smooth bounded open set, $N \geq 3$ and $2 < p < 2^* = 2N/(N-2)$, then (9.17) admits an unbounded sequence (u_k) of solutions $u_k \in W_0^{1,2}(\Omega)$. Here we shall only localize one of the solutions, in a shell.

In context we shall obtain an upper bound for the quantity

$$\lambda_{p-1} = \inf \left\{ \frac{\int_\Omega |u|^{p-2} |\nabla u|^2 \, dx}{\left(\int_\Omega |u|^{p^2/2} \, dx\right)^{2/p}} : \begin{array}{l} u \in C^1\left(\overline{\Omega}\right) \setminus \{0\}, \\ u = 0 \text{ on } \partial\Omega \end{array} \right\}. \tag{9.18}$$

For $p = 2$, λ_1 is the *first eigenvalue* of the *Laplacian* $-\Delta$ under the Dirichlet boundary condition, and $1/\lambda_1$ represents the best constant in the Poincaré's inequality (see Precup [30], Theorem 3.59). For $p > 2$ and $N = 1$ this quantity arises in the study of compactness properties of integral operators on spaces of functions with values in a Banach space (see Precup [33]).

The topics in this section require a previous knowledge of linear partial differential equations.

We now recall some well known results from the theory of linear elliptic boundary value problems.

(P1) Let $\Omega \subset \mathbf{R}^N$ be a bounded open set with C^2-boundary. The Laplacian $-\Delta$ is a self-adjoint operator on $L^2(\Omega)$ with domain $W^{2,2}(\Omega) \cap W_0^{1,2}(\Omega)$ (see Precup [30], Theorem 3.33). It can be regarded as a continuous operator from $W^{2,q}(\Omega) \cap W_0^{1,q}(\Omega)$ to $L^q(\Omega)$ for each $q \in (1, \infty)$. Moreover, $-\Delta$ is invertible and

$$A := (-\Delta)^{-1}$$

is a bounded linear operator from $L^q(\Omega)$ to $W^{2,q}(\Omega)$ (see Brezis [5], Theorem 9.32). Also, A as an operator from $L^2(\Omega)$ to $L^2(\Omega)$ is positive, self-adjoint, and completely continuous.

(P2) (Sobolev embedding theorem) Let $\Omega \subset \mathbf{R}^N$ be a bounded open set with Lipschitz boundary, $k \in \mathbf{N}$, $1 \leq q \leq \infty$. Then the following holds:

(1^0) If $kq < N$ we have

$$W^{k,q}(\Omega) \subset L^r(\Omega) \tag{9.19}$$

and the embedding is continuous for $r \in [1, Nq/(N - kq)]$; the embedding is compact if $r \in [1, Nq/(N - kq))$.

(2^0) If $kq = N$, then (9.19) holds with compact embedding for $r \in [1, \infty)$.

(3^0) If $0 \leq m < k - N/q < m + 1$ we have

$$W^{k,q}(\Omega) \subset C^{m,\alpha}(\overline{\Omega}) \tag{9.20}$$

and the embedding is continuous for $0 \leq \alpha \leq k - m - N/q$; the embedding is compact if $\alpha < k - m - N/q$.

The above results are valid for $W_0^{k,q}(\Omega)$ spaces on arbitrary bounded domains Ω (see Adams [1], Struwe [42] p 213, or Brezis [5] pp 168–169).

(P3) Let $\Omega \subset \mathbf{R}^N$ be a bounded open set with C^2-boundary. Let $p_0 = 2N/(N-2)$ if $N \geq 3$ and p_0 be any number of $(2, \infty)$ if $N = 1$ or $N = 2$. Let q_0 be the conjugate of p_0. Clearly, $p_0 \in (2, \infty)$ and $q_0 \in (1, 2)$. From (P1), (P2) we have that A has the following properties:

(a) $A : L^q(\Omega) \to L^p(\Omega)$ for every $q \in [q_0, 2]$, $1/p + 1/q = 1$;

(b) A is continuous from $L^q(\Omega)$ to $L^p(\Omega)$ for every $q \in [q_0, 2]$, $1/p + 1/q = 1$;

(c) as a map from $L^2(\Omega)$ to $L^2(\Omega)$, A is a positive self-adjoint operator.

Indeed, A is continuous from $L^q(\Omega)$ to $W^{2,q}(\Omega)$. On the other hand

$$W^{2,q}(\Omega) \subset L^p(\Omega)$$

with continuous embedding. This is clear if $q \geq N/2$. For $q < N/2$ and $1/p + 1/q = 1$, observe that

$$p \leq \frac{2N}{N-2} \Longleftrightarrow p \leq \frac{Nq}{N-2q}.$$

According to Theorem 6.2 the properties (a)–(c) are sufficient for that the operator A considered from $L^q(\Omega)$ to $L^p(\Omega)$, where $p \in (2, p_0)$ and $1/p + 1/q = 1$, admits a representation in the form

$$A = HH^*$$

where

$$H : L^2(\Omega) \to L^p(\Omega), \quad H(v) = A^{1/2}(v)$$

and

$$H^* : L^q(\Omega) \to L^2(\Omega)$$

is the adjoint of H. Notice H is injective since A is positive defined.

In what follows by $|H|$ we shall mean

$$|H| = \sup\{|H(v)|_p : v \in L^2(\Omega), \ |v|_2 = 1\}.$$

Our first result gives a lower bound for all nontrivial solutions of (9.17).

Theorem 9.5 *Let Ω be a bounded open set of $\mathbf{R}^N$ with C^2-boundary and let $p \in (2, 2N/(N-2))$ if $N \geq 3$, $p \in (2, \infty)$ if $N = 1$ or $N = 2$, and let q be the conjugate of p. If $u \in W^{2,q}(\Omega) \cap W_0^{1,q}(\Omega)$ is a non-zero solution of (9.17), then the function*

$$v = H^*\left(|u|^{p-2}u\right) = H^{-1}(u)$$

satisfies

$$|v|_2 \geq |H|^{-1}\left[(p-1)\lambda_{p-1}\right]^{1/(p-2)}. \tag{9.21}$$

Proof. Let us first prove that any solution of (9.17) belongs to $C^1(\overline{\Omega})$.

For $N = 1$ this follows from (9.20) (choose $\alpha = 0$, $m = 1$ and $k = 2$). Assume $N \geq 2$ and fix any number $q_0 > N(p-1)$. If $q \geq N/2$ then (P2) guarantees $u \in L^{q_0}(\Omega)$. Assume $q < N/2$ and denote $q_1 = q$. Since $u \in W^{2,q_1}(\Omega)$ and $q_1 < N/2$, from (9.19) we have $u \in L^{q_1^*}(\Omega)$, where

$q_1^* = Nq_1/(N-2q_1)$. Then $|u|^{p-2}u \in L^{q_1^*/(p-1)}(\Omega)$. Let $q_2 = q_1^*/(p-1)$. Since $u = A\left(|u|^{p-2}u\right)$ and $|u|^{p-2}u \in L^{q_2}(\Omega)$, from (P1) we have that $u \in W^{2,q_2}(\Omega)$. If $q_2 \geq N/2$, as above $u \in L^{q_0}(\Omega)$; otherwise we continue this way. At the step j we find that

$$u \in W^{2,q_j}(\Omega), \quad q_j = \frac{q_{j-1}^*}{p-1}, \quad q_{j-1}^* = \frac{Nq_{j-1}}{N-2q_{j-1}}, \tag{9.22}$$

where $q_1, q_2, ..., q_{j-1} < N/2$ $(j \geq 2)$. We claim that there exists a j with $q_j \geq N/2$. To prove this, assume the contrary, that is $q_j < N/2$ for every $j \geq 1$. Using $p < 2N/(N-2)$ we can show by induction that the sequence (q_j) is increasing. Consequently, $q_j \to \bar{q} \in [q, N/2]$ as $j \to \infty$. Next, from (9.22) we obtain

$$q_j(N-2q_{j-1})(p-1) = Nq_{j-1}.$$

Letting $j \to \infty$ this yields $\bar{q}(N-2\bar{q})(p-1) = N\bar{q}$ and so

$$\begin{aligned}\bar{q} &= \frac{N(p-2)}{2(p-1)} \\ &\geq q = \frac{p}{p-1}.\end{aligned}$$

This implies $p \geq 2N/(N-2)$, a contradiction. Thus our claim is proved. Therefore $u \in L^{q_0}(\Omega)$. Furthermore, $|u|^{p-2}u \in L^{q_0/(p-1)}(\Omega)$, and since $u = A\left(|u|^{p-2}u\right)$ we have $u \in W^{2,q_0/(p-1)}(\Omega)$. Since $q_0/(p-1) > N$, by (9.20) one has $W^{2,q_0/(p-1)}(\Omega) \subset C^1(\overline{\Omega})$ (choose $\alpha = 0$, $k = 2$, $m = 1$). Hence $u \in C^1(\overline{\Omega})$ as claimed.

Let $\overline{u} = A\left(|u|^{p-1}\right)$, where u is a non-zero solution of (9.17) and let $v = H^{-1}(u)$. Clearly, like u, $\overline{u} \in C^1(\overline{\Omega})$ and $\overline{u} = 0$ on $\partial\Omega$. By the weak maximum principle we have

$$|u| \leq \overline{u} \text{ on } \overline{\Omega}. \tag{9.23}$$

Hence

$$-\Delta\overline{u} = |u|^{p-1} \leq |u|^{p-2}\overline{u}.$$

If we 'multiply' by $\overline{u}^{p-1}$ and 'integrate' on Ω, we obtain

$$(p-1)\int_\Omega \overline{u}^{p-2}|\nabla\overline{u}|^2\,dx \leq \int_\Omega |u|^{p-2}\,\overline{u}^p dx. \tag{9.24}$$

Now, Hölder's inequality yields

$$\begin{aligned}\int_\Omega |u|^{p-2}\,\overline{u}^p dx &\leq \left(\int_\Omega \overline{u}^{p^2/2} dx\right)^{2/p} \left(\int_\Omega |u|^p\, dx\right)^{(p-2)/p} \qquad (9.25)\\ &= |H(v)|_p^{p-2} \left(\int_\Omega \overline{u}^{p^2/2} dx\right)^{2/p}.\end{aligned}$$

Since $|H(v)|_p \leq |H|\,|v|_2$ and by (9.23) one has $\overline{u} \neq 0$, from (9.24) and (9.25) we deduce that

$$(p-1)\,\lambda_{p-1} \leq |H|^{p-2}\,|v|_2^{p-2},$$

that is, (9.21). ■

The main result in this section is the following existence and localization theorem.

Theorem 9.6 *Let Ω be a bounded open set of $\mathbf{R}^N$ with C^2-boundary and let $p \in (2, 2N/(N-2))$ if $N \geq 3$, and $p \in (2, \infty)$ if $N = 1$ or $N = 2$. Then the problem (9.17) has a solution u with*

$$|H|^{-1}\left[(p-1)\,\lambda_{p-1}\right]^{1/(p-2)} \leq \left|H^{-1}(u)\right|_2 \leq |H|^{-p/(p-2)}. \qquad (9.26)$$

Proof. First notice the left inequality in (9.26) is true for all non-zero solutions of (9.17) according to Theorem 9.5. Next we prove that for each

$$R > r = |H|^{-p/(p-2)},$$

(9.17) has a solution u such that

$$\left|H^{-1}(u)\right|_2 \leq R. \qquad (9.27)$$

The equation (9.17) can be put under the form (9.1), where $N = N_f$ and

$$f(x,z) = |z|^{p-2}\,z \quad (x \in \Omega,\ z \in \mathbf{R}).$$

Also

$$F(x,z) = \frac{1}{p}\,|z|^p.$$

Thus the Golomb functional $E : L^2(\Omega) \to \mathbf{R}$ is given by

$$\begin{aligned}E(v) &= \frac{1}{2}\,|v|_2^2 - \frac{1}{p}\int_\Omega |H(v)|^p\, dx\\ &= \frac{1}{2}\,|v|_2^2 - \frac{1}{p}\,|H(v)|_p^p.\end{aligned}$$

As in the proof of Theorem 9.3 we can show that E satisfies the $(\mathrm{PS})_R$ condition.

If for this R the Leray–Schauder boundary condition (9.7) does not hold, then there are $v \in L^2(\Omega)$ and $\lambda \in (0,1)$ such that $|v|_2 = R$ and

$$v = \lambda H^* N H(v).$$

It is easy to see that the function $\overline{v} = \lambda^{1/(p-2)} v$ satisfies

$$\overline{v} = H^* N H(\overline{v}),$$

i.e., $\overline{v}$ is a critical point of E. Clearly

$$0 < |\overline{v}|_2 < |v|_2 = R.$$

Then $u = H(\overline{v})$ is a solution of (9.17) and satisfies (9.27). Assume now that (9.7) holds for this R. We claim that the energy inequality (9.6) also holds with a suitable $v_1 \in L^2(\Omega)$ satisfying $r < |v_1|_2 \le R$. To prove this we fix any $\lambda \in (r, R]$ and we choose a $\rho > 0$ sufficiently small so that

$$\phi(\lambda) + \frac{1}{p}\lambda^p \rho < \phi(r), \tag{9.28}$$

where

$$\phi(\sigma) = \frac{1}{2}\sigma^2 - \frac{1}{p}\sigma^p |H|^p, \quad \sigma \in \mathbf{R}_+.$$

Notice r is the maximum point of ϕ, ϕ is increasing on $[0, r]$ and decreasing on $[r, \infty)$. Now we choose a function $v_2 \in L^2(\Omega)$ with $|v_2|_2 = 1$ and

$$|H(v_2)|_p^p \ge |H|^p - \rho.$$

Then (9.6) holds for $v_1 = \lambda v_2$. Indeed

$$\begin{aligned} E(v_1) &= E(\lambda v_2) = \frac{1}{2}\lambda^2 - \frac{1}{p}\lambda^p |H(v_2)|_p^p \\ &\le \frac{1}{2}\lambda^2 - \frac{1}{p}\lambda^p |H|^p + \frac{1}{p}\lambda^p \rho = \phi(\lambda) + \frac{1}{p}\lambda^p \rho. \end{aligned} \tag{9.29}$$

Also, for every $v \in L^2(\Omega)$ with $|v|_2 = r$ we have

$$\begin{aligned} E(v) &= \frac{1}{2}r^2 - \frac{1}{p}r^p \left|H(r^{-1}v)\right|_p^p \\ &\ge \frac{1}{2}r^2 - \frac{1}{p}r^p |H|^p = \phi(r). \end{aligned} \tag{9.30}$$

Now (9.29), (9.30) and (9.28) guarantee (9.6). Thus all the assumptions of Theorem 9.1 hold and so there exists a non-zero critical point v of E with $|v|_2 \leq R$. Therefore for each $R > r$ (9.17) has a solution $u \in L^p(\Omega)$ satisfying (9.27).

Finally, for each positive integer k we put $R = r + 1/k$ to obtain a solution u_k with

$$\left|H^{-1}(u_k)\right|_2 \leq r + \frac{1}{k},$$

and the result will follow via a limit argument. ■

Remark 9.4 From (9.26) we have the inequality:

$$\lambda_{p-1} \leq \frac{1}{(p-1)\,|H|^2}.$$

The next inequality of Poincaré type shows that

$$\lambda_{p-1} > 0$$

for $p \in [2, 2N/(N-2)]$ if $N \geq 3$ and for $p \in [2, \infty)$ if $N = 2$. Moreover, its proof connects λ_{p-1} to the embedding constant of $W_0^{1,2}(\Omega)$ into $L^p(\Omega)$.

Theorem 9.7 *Let $\Omega \subset \mathbf{R}^N$ be bounded open and let $p \in [2, 2N/(N-2)]$ if $N \geq 3$, and $p \in [2, \infty)$ if $N = 2$. Then there exists a constant $c > 0$ depending only on p and Ω, such that*

$$\left(\int_\Omega |u|^{p^2/2}\,dx\right)^{2/p} \leq c \int_\Omega |u|^{p-2}\,|\nabla u|^2\,dx \tag{9.31}$$

for all $u \in C^1(\overline{\Omega})$ with $u = 0$ on $\partial\Omega$.

Proof. According to (P2), we have $W_0^{1,2}(\Omega) \subset L^p(\Omega)$ with continuous embedding. Hence there exists a constant $c_0 > 0$ such that

$$|v|_p \leq c_0\,|v|_{W_0^{1,2}(\Omega)} \quad \text{for all } v \in W_0^{1,2}(\Omega).$$

Here

$$|v|_{W_0^{1,2}(\Omega)} = \left(\int_\Omega |\nabla v|^2\,dx\right)^{1/2}.$$

Since $C_0^\infty(\Omega)$ is dense in $W_0^{1,2}(\Omega)$ we may assume that

$$c_0 = \sup\left\{|v|_p : v \in C_0^\infty(\Omega),\ |v|_{W_0^{1,2}(\Omega)} = 1\right\}.$$

The space $C_0^\infty(\Omega)$ is also dense in $\{u \in C^1(\overline{\Omega}) : u = 0 \text{ on } \partial\Omega\}$, and so

$$\lambda_{p-1} = \left(\sup\left\{\left(\int_\Omega |u|^{p^2/2}\, dx\right)^{2/p} : u \in C_0^\infty(\Omega),\ \int_\Omega |u|^{p-2} |\nabla u|^2\, dx = 1\right\}\right)^{-1}.$$

After substituting $v = (2/p)\,|u|^{p/2}$ we obtain

$$\begin{aligned}\lambda_{p-1} &= \left(\frac{2}{p}\right)^2 \left(\sup\left\{|v|_p^2 : v \in C_0^\infty(\Omega),\ |v|_{W_0^{1,2}(\Omega)} = 1\right\}\right)^{-1} \\ &= \left(\frac{2}{p\,c_0}\right)^2.\end{aligned}$$

Thus (9.31) holds with the smallest constant

$$c = \frac{1}{\lambda_{p-1}} = \left(\frac{p\,c_0}{2}\right)^2,$$

and the proof is complete. ■

For other applications of variational techniques to integral equations and boundary value problems for various types of equations, see Conti–Merizzi–Terracini [9], Dincă–Jebelean–Mawhin [11], Mawhin–Willem [23], Moroz–Vignoli–Zabreiko [24], Moroz–Zabreiko [25], [26], Pascali [28], Precup [29], Rabinowitz [36], Rădulescu [37], Ricceri [38], Struwe [42] and Zeidler [45].

References: Part II

[1] ADAMS, R.A., *Sobolev Spaces,* Academic Press, London, 1978.

[2] AMBROSETTI, A., Variational methods and nonlinear problems: classical results and recent advances, In: *Topological Nonlinear Analysis* (M. Matzeu and A. Vignoli eds.), Birkhäuser, Boston–Basel–Berlin, 1995, 1–36.

[3] AMBROSETTI, A. and RABINOWITZ, P.H., Dual variational methods in critical point theory and applications, *J. Funct. Anal.* **14** (1973), 349–381.

[4] BENCI, V., Some critical point theorems and applications, *Comm. Pure Appl. Math.* **33** (1980), 147–172.

[5] BREZIS, H., *Analyse fonctionnelle,* Masson, Paris, 1983.

[6] BREZIS, H. and NIRENBERG, L., Remarks on finding critical points, *Comm. Pure Appl. Math.* **44** (1991), 939–963.

[7] CANINO, A. and DEGIOVANNI, M., Nonsmooth critical point theory and quasilinear elliptic equations. In: *Topological Methods in Differential Equations and Inclusions* (A. Granas and M. Frigon eds.), Kluwer Academic Publ., Dordrecht–Boston–London, 1995, 1–50.

[8] CHANG, K.C., Variational methods for non-differentiable functionals and their applications to partial differential equations, *J. Math. Anal. Appl.* **80** (1981), 102–129.

[9] CONTI, M., MERIZZI, L. and TERRACINI, S., Remarks on variational methods and lower–upper solutions, *NoDEA Nonlinear Differential Equations Appl.* **4** (1999), 371–393.

[10] CORVELLEC, J.-N., Quantitative deformation theorems and critical point theory, *Pacific J. Math.* **187** (1999), 263–279.

[11] DINCĂ, G., JEBELEAN, P. and MAWHIN, J., A result of Ambrosetti–Rabinowitz type for p-Laplacian. In: *Qualitative Problems for Differential Equations and Control Theory* (C. Corduneanu ed.), World Scientific, Singapore–New Jersey–London–Hong Kong, 1995, 231–243.

[12] EKELAND, I., On the variational principle, *J. Math. Anal. Appl.* **47** (1974), 324–353.

[13] FRIGON, M., On a new notion of linking and application to elliptic problems at resonance, *J. Differential Equations* **153** (1999), 96–120.

[14] GHOUSSOUB, N. and PREISS, D., A general mountain pass principle for locating and classifying critical points, *Ann. Inst. Poincaré Anal. Nonlinéaire* **6** (1989), 321–330.

[15] GOLOMB, M., Zur Theory der nichtlinearen Integralgleichungen, Integralgleichungssysteme und algemeinen Functionalglichungen, *Math. Zeit.* **39** (1934), 45–75.

[16] GUO, D., SUN, J. and QI, G., Some extensions of the mountain pass lemma, *Differential Integral Equations* **1** (1988), 351–358..

[17] HAMMERSTEIN, A., Nichtlineare Integralgleichungen nebst Anwendungen, *Acta Math.* **54** (1930), 117–176.

[18] KAVIAN, O., *Introduction à la théorie des points critiques et applications aux problèmes elliptiques,* Springer–Verlag, Paris, 1993.

[19] KRASNOSELSKII, M.A., *Topological Methods in the Theory of Nonlinear Integral Equations,* Pergamon Press, Oxford–London–New York–Paris, 1964.

[20] KRASNOSELSKII, M.A., ZABREIKO, P.P., PUSTYLNIK, E.I. and SOBOLEVSKII, P.E., *Integral Operators in Spaces of Summable Functions,* Noordhoff Publ., Leyden, 1976.

[21] LIU, Z. and SUN, J., Invariant sets of descending flow in critical point theory with applications to nonlinear differential equations, *J. Differential Equations* **172** (2001), 257–299.

[22] MA, L., Mountain pass on a closed convex set, *J. Math. Anal. Appl.* **205** (1997), 531–536.

[23] MAWHIN, J. and WILLEM, M., *Critical Point Theory and Hamiltonian Systems,* Springer–Verlag, New York–Berlin–Heidelberg–London–Paris–Tokyo, *1989.*

[24] MOROZ, V., VIGNOLI, A. and ZABREIKO, P., On the three critical points theorem, *Topol. Methods Nonlinear Anal.* **11** (1998), 103–113.

[25] MOROZ, V. and ZABREIKO, P., A variant of the mountain pass theorem and its application to Hammerstein integral equations, *Z. Anal. Anwendungen* **15** (1996), 985–997.

[26] MOROZ, V. and ZABREIKO, P., On the Hammerstein equations with natural growth conditions, *Z. Anal. Anwendungen* **18** (1999), 625–638.

[27] MOTREANU, D. and VARGA, C., A nonsmooth equivariant minimax principle, *Commun. Appl. Anal.* **3** (1999), 115–130.

[28] PASCALI, D., Some mountain climbing techniques in the theory of semilinear operator equations, *Libertas Math.* **13** (1993), 155–170.

[29] PRECUP, R., *Nonlinear Integral Equations* (Romanian), Babeş–Bolyai University, Cluj, 1993.

[30] PRECUP, R., *Partial Differential Equations* (Romanian), Transilvania Press, Cluj, 1997.

[31] PRECUP, R., Nontrivial solvability of Hammerstein integral equations in Hilbert spaces. In: *Séminaire de la Théorie de la Meilleure Approximation, Convexité et Optimisation* (E. Popoviciu ed.), Srima, Cluj–Napoca, 2000, 255–265.

[32] PRECUP, R., On the Palais–Smale condition for Hammerstein integral equations in Hilbert spaces, *Nonlinear Anal.* **47** (2001), 1233–1244.

[33] PRECUP, R., Inequalities and compactness. In: *Inequality Theory and Applications, Volume 1* (Y.J. Cho, J.K. Kim and S.S. Dragomir eds.), Nova Science Publishers, Huntington–New York, 2001, 257–271.

[34] PRECUP, R., An inequality which arises in the absence of the mountain pass geometry, *J. Inequal. Pure Appl. Math.* **3** (2002), no. 3, 1–10.

[35] PUCCI, P. and SERRIN, J., A mountain pass theorem, *J. Differential Equations* **60** (1985), 142–149.

[36] RABINOWITZ, P.H., *Minimax Methods in Critical Point Theory with Applications to Differential Equations,* CBMS Regional Conference Series Math., Vol. 65, Amer. Math. Soc., Providence, 1986.

[37] RĂDULESCU, V., *Treatment Methods of the Elliptic Problems,* Univ. of Craiova, Craiova, 1998.

[38] RICCERI, B., A general variational principle and some of its applications, *J. Comput. Appl. Math.* **113** (2000), 401–410.

[39] SCHECHTER, M., A bounded mountain pass lemma without the (PS) condition and applications, *Trans. Amer. Math. Soc.* **330** (1992), 681–703.

[40] SCHECHTER, M., *Linking Methods in Critical Point Theory,* Birkhäuser, Boston–Basel–Berlin, 1999.

[41] SCHECHTER, M. and TINTAREV, K., Nonlinear eigenvalues and mountain pass methods, *Topol. Methods Nonlinear Anal.* **1** (1993), 183–201.

[42] STRUWE, M., *Variational Methods,* Springer–Verlag, Berlin–Heidelberg–New York–London–Paris–Tokyo–Hong Kong–Barcelona, 1990.

[43] VAINBERG, M.M., *Variational Methods for the Study of Nonlinear Operators,* Holden–Day, San Francisco, 1964.

[44] YOSIDA, K., *Functional Analysis,* Springer–Verlag, Berlin, 1974.

[45] ZEIDLER, E., *Nonlinear Functional Analysis and its Applications, Vol.* III, Springer–Verlag, New York–Berlin–Heidelberg–Tokyo, 1985.

Part III

ITERATIVE METHODS

Chapter 10

The Discrete Continuation Principle

The Banach contraction principle was generalized by Perov (see Perov–Kibenko [40]) for contractive maps on spaces endowed with vector-valued metrics. Also, Granas [25] proved that the property of having a fixed point is invariant under homotopy for contractions on complete metric spaces. This result was completed in Precup [50] (see also O'Regan–Precup [38] and Precup [51], [52]) by an iterative procedure of discrete continuation along the fixed points curve. This chapter presents a variant for contractive maps on spaces with vector-valued metrics, first given in Precup [54].

10.1 Perov's Theorem

Let X be a nonempty set. By a *vector-valued metric* on X we mean a map $d : X \times X \to \mathbf{R}^n$ with the following properties:

(i) $d(u,v) \geq 0$ for all $u,\ v \in X$; if $d(u,v) = 0$ then $u = v$;
(ii) $d(u,v) = d(v,u)$ for all $u,\ v \in X$;
(iii) $d(u,v) \leq d(u,w) + d(w,v)$ for all $u,\ v,\ w \in X$.

Here, if $x,\ y \in \mathbf{R}^n$, $x = (x_1, x_2, ..., x_n)$, $y = (y_1, y_2, ..., y_n)$, and $c \in \mathbf{R}$, by $x \leq y$ we mean that $x_i \leq y_i$ for $i = 1, 2, ..., n$, whilst by $x \leq c$ we mean that $x_i \leq c$ for all i. Also $x < y$ stands for $x_i < y_i$, $i = 1, 2, ..., n$, and $x < c$ for $x_i < c$, $i = 1, 2, ..., n$. For a vector $x = (x_1, x_2, ..., x_n)$ we write $x = \infty$ if $x_i = \infty$ for all i.

A set X endowed with a vector-valued metric d is said to be a *generalized metric space*. For the generalized metric spaces the notions of a convergent sequence, Cauchy sequence, completeness, open subset, and closed subset are similar to those for usual metric spaces.

Definition 10.1 Let (X, d) be a generalized metric space. A map $T : X \to X$ is said to be *contractive* if there exists a matrix $M \in M_{n \times n}(\mathbf{R}_+)$ such that

$$M^k \to 0 \quad \text{as} \quad k \to \infty \tag{10.1}$$

and

$$d(T(u), T(v)) \leq M\, d(u, v) \tag{10.2}$$

for all $u, v \in X$. A matrix M which satisfies (10.1) is said to be *convergent to zero.*

Lemma 10.1 *Let $M \in \mathbf{M}_{n\times n}(\mathbf{R}_+)$. The matrix M is convergent to zero if and only if $I - M$ is non-singular and*

$$(I - M)^{-1} = I + M + M^2 + \ldots . \tag{10.3}$$

Here I is the unit matrix of order n.

Proof. Assume first that M is convergent to zero. To show that $I - M$ is non-singular it suffices to prove that the linear system

$$(I - M)\, z = 0$$

has only the null solution. Assume $z \in \mathbf{R}^n$ is a solution of this system. Then

$$z = Mz = M^2 z = \ldots = M^k z = \ldots$$

and letting $k \to \infty$ we deduce $z = 0$. Hence $I - M$ is non-singular. Furthermore, (10.3) follows from the identity

$$I - (I - M)\left(I + M + M^2 + \ldots + M^k\right) = M^{k+1}. \tag{10.4}$$

Conversely, if $I - M$ is non-singular and (10.3) holds, then from the convergence of the series (10.3) we conclude that $M^k \to 0$ as $k \to \infty$. ■

Remark 10.1 Property (10.1) is also equivalent to the property of the eigenvalues of M that they are all located inside the unit disc of the complex plane (see Rus [55], Theorem 4.1.1).

Remark 10.2 From (10.3) we have that

$$z \le (I - M)^{-1} z$$

for every $z \in R_+^n$. In particular, the set

$$P_M = \{z \in \mathbf{R}^n : z > 0 \text{ and } (I - M) z > 0\}$$

is nonempty and coincides with the set

$$\left\{ (I - M)^{-1} z_0 : z_0 \in \mathbf{R}^n, \ z_0 > 0 \right\}.$$

Indeed, if $z_0 \in R^n$ and $z_0 > 0$ then

$$z := (I - M)^{-1} z_0 \ge z_0,$$

and so $z > 0$. In addition $(I - M) z = z_0$. Hence $(I - M) z > 0$ and so $z \in P_M$. Conversely, if $z \in P_M$ then $z_0 := (I - M) z > 0$ and $z = (I - M)^{-1} z_0$.

Remark 10.3 Assume that M is a matrix convergent to zero. A subset $U \subset X$ is open if and only if for each $u \in U$ there is a $\rho \in P_M$ such that $v \in U$ whenever $d(v, u) \le \rho$ (take $\rho = \varepsilon \rho_0$ for any $\rho_0 \in P_M$ and $\varepsilon > 0$ sufficiently small).

Theorem 10.1 (Perov) *Let (X, d) be a complete generalized metric space and $T : X \to X$ be contractive with the Lipschitz matrix M. Then T has a unique fixed point u^*, and for each $u_0 \in X$ one has*

$$d\left(T^k(u_0), u^*\right) \le M^k (I - M)^{-1} d(u_0, T(u_0)) \tag{10.5}$$

for all $k \in \mathbf{N}$.

Proof. Let u_0 be any element of X. Define a sequence (u_k) by

$$u_{k+1} = T(u_k), \quad k \in \mathbf{N}. \tag{10.6}$$

Then from (10.2)

$$d(u_k, u_{k+1}) \le M^k d(u_0, u_1)$$

and, as a consequence,

$$d(u_k, u_{k+m}) \le \left(M^k + M^{k+1} + ... + M^{k+m-1}\right) d(u_0, u_1).$$

Using (10.3) we deduce that

$$d(u_k, u_{k+m}) \leq M^k (I - M)^{-1} d(u_0, u_1). \tag{10.7}$$

Hence (u_k) is a Cauchy sequence. Let u^* be its limit. Then from (10.6) we have $u^* = T(u^*)$, whilst from (10.7) and letting $m \to \infty$ we obtain (10.5). For uniqueness let u_1, u_2 be two fixed points of T, then

$$d(u_1, u_2) = d\left(T^k(u_1), T^k(u_2)\right) \leq M^k d(u_1, u_2).$$

Since $M^k \to 0$ as $k \to \infty$, this implies $d(u_1, u_2) = 0$, so $u_1 = u_2$. ■

For some applications of Perov's theorem to systems of integral equations, see Avramescu [3], Perov–Kibenko [40] and Rus [55].

10.2 The Continuation Principle for Contractive Maps on Generalized Metric Spaces

The main abstract result of this chapter is the following theorem from Precup [54].

Theorem 10.2 *Let (X, d) be a complete generalized metric space where $d: X \times X \to \mathbf{R}^n$ and let U be an open set of X. Let $H : \overline{U} \times [0,1] \to X$ and assume that the following conditions are satisfied:*

(a1) there is a matrix $M \in \mathbf{M}_{n \times n}(\mathbf{R}_+)$ convergent to zero such that

$$d(H(u, \lambda), H(v, \lambda)) \leq M\, d(u, v) \tag{10.8}$$

for all $u, v \in \overline{U}$ and $\lambda \in [0,1]$;

(a2) $H(u, \lambda) \neq u$ for all $u \in \partial U$ and $\lambda \in [0,1]$;

(a3) H is continuous in λ, uniformly for $u \in \overline{U}$, i.e., for each $\varepsilon \in (0, \infty)$ and $\lambda \in [0,1]$ there is a $\rho \in (0, \infty)$ such that

$$d(H(u, \lambda), H(u, \mu)) < \varepsilon$$

whenever $u \in \overline{U}$ and $|\lambda - \mu| < \rho$.

In addition assume that $H_0 := H(.,0)$ has a fixed point. Then for each $\lambda \in [0,1]$ there exists a unique fixed point $u(\lambda)$ of $H_\lambda := H(., \lambda)$. Moreover, $u(\lambda)$ depends continuously on λ and there exists $r \in P_M \cup \{\infty\}$, integers m, k_1, k_2, ..., k_{m-1} and numbers

$$0 < \lambda_1 < \lambda_2 < \ldots < \lambda_{m-1} < \lambda_m = 1$$

such that for any $u_0 \in X$ *satisfying* $d(u_0, u(0)) \leq r$, *the sequences* $(u_{j,i})_{i\geq 0}$, $j = 1, 2, ..., m$, *defined by*

$$\begin{aligned} u_{1,0} &= u_0 \\ u_{j,i+1} &= H_{\lambda_j}(u_{j,i}), \quad i = 0, 1, ... \\ u_{j+1,0} &= u_{j,k_j}, \qquad j = 1, 2, ..., m-1 \end{aligned}$$

are well defined and satisfy

$$d(u_{j,i},\, u(\lambda_j)) \leq M^i (I - M)^{-1} d(u_{j,0},\, H_{\lambda_j}(u_{j,0})), \quad i \in \mathbf{N}. \tag{10.9}$$

Proof. *Step* 1. First we prove that for each $\lambda \in [0,1]$ H_λ has a fixed point. Let

$$\Lambda = \{\lambda \in [0,1]\,;\ H(u,\lambda) = u \text{ for some } u \in U\}.$$

We have $0 \in \Lambda$ by the assumption that H_0 has a fixed point. Hence Λ is nonempty. We will show that Λ is both closed and open in $[0,1]$ and so, by the connectedness of $[0,1]$, $\Lambda = [0,1]$.

To prove that Λ is closed let $\lambda_k \in \Lambda$ with $\lambda_k \to \lambda$ as $k \to \infty$. Since $\lambda_k \in \Lambda$ there is $u_k \in U$ so that $H(u_k, \lambda_k) = u_k$. Then by (10.8) we obtain

$$\begin{aligned} d(u_k, u_j) &= d(H(u_k, \lambda_k), H(u_j, \lambda_j)) \\ &\leq d(H(u_k, \lambda_k), H(u_k, \lambda)) \\ &\quad + d(H(u_k, \lambda), H(u_j, \lambda)) + d(H(u_j, \lambda), H(u_j, \lambda_j)) \\ &\leq d(H(u_k, \lambda_k), H(u_k, \lambda)) + M\, d(u_k, u_j) \\ &\quad + d(H(u_j, \lambda), H(u_j, \lambda_j)). \end{aligned}$$

It follows that

$$d(u_k, u_j) \leq (I - M)^{-1} \left[d(H(u_k, \lambda_k), H(u_k, \lambda)) + d(H(u_j, \lambda), H(u_j, \lambda_j)) \right].$$

This shows by (a3) that (u_k) is a Cauchy sequence. Then since X is complete there is a $u \in X$ with $d(u_k, u) \to 0$ as $k \to \infty$. Clearly $u \in \overline{U}$. Then

$$d(u_k, H(u, \lambda)) \to d(u, H(u, \lambda))$$

and, from (10.8) and (a3), we have

$$\begin{aligned} d(u_k, H(u, \lambda)) &= d(H(u_k, \lambda_k), H(u, \lambda)) \\ &\leq d(H(u_k, \lambda_k), H(u, \lambda_k)) + d(H(u, \lambda_k), H(u, \lambda)) \\ &\leq M d(u_k, u) + d(H(u, \lambda_k), H(u, \lambda)) \to 0. \end{aligned}$$

Hence $d(u, H(u, \lambda)) = 0$, that is $H(u, \lambda) = u$. By (a2) $u \in U$, and so $\lambda \in \Lambda$.

To prove that Λ is open in $[0,1]$ let $\mu \in \Lambda$ and $w \in U$ such that $H(w, \mu) = w$. Since U is open there exists a $\rho \in P_M$ (see Remark 10.3) such that

$$d(u, w) \leq \rho \quad \text{implies} \quad u \in U.$$

Also, by (a3) there is $\eta = \eta(\rho) \in (0, \infty)$ such that

$$d(w, H(w, \lambda)) = d(H(w, \mu), H(w, \lambda)) \leq (I - M)\rho \tag{10.10}$$

for $|\lambda - \mu| \leq \eta$. Consequently

$$\begin{aligned} d(w, H(u, \lambda)) &\leq d(w, H(w, \lambda)) + d(H(w, \lambda), H(u, \lambda)) \\ &\leq (I - M)\rho + Md(w, u) \leq \rho \end{aligned}$$

whenever $d(w, u) \leq \rho$ and $|\lambda - \mu| \leq \eta$. This shows that for $|\lambda - \mu| \leq \eta$, H_λ sends B into itself, where $B = \{u \in X : d(w, u) \leq \rho\}$. Notice (B, d) is a complete generalized metric space. Hence for $|\lambda - \mu| \leq \eta$, by Theorem 10.1 H_λ has a fixed point $u(\lambda) \in B \subset U$. This shows that μ is an interior point of Λ and hence Λ is open in $[0,1]$. Notice for every $u \in B$ and $|\lambda - \mu| \leq \eta$, we also have by Theorem 10.1 that the sequence $(H_\lambda^k(u))_{k \geq 0}$ is well defined and

$$d(H_\lambda^k(u), u(\lambda)) \leq M^k (I - M)^{-1} d(u, H_\lambda(u)), \quad k \in \mathbf{N}.$$

Step 2. The uniqueness of $u(\lambda)$ is a consequence of (10.8). Indeed, if $u, v \in U$ are fixed points of H_λ then since the elements of M are nonnegative, we have

$$\begin{aligned} d(u, v) &= d(H(u, \lambda), H(v, \lambda)) \\ &\leq Md(u, v) \\ &\leq M^2 d(u, v) \\ &\vdots \\ &\leq M^k d(u, v) \to 0 \quad \text{as } k \to \infty. \end{aligned}$$

Hence $d(u, v) = 0$, and so $u = v$.

Step 3. $u(\lambda)$ is continuous on $[0,1]$. Indeed,

$$\begin{aligned} & d(u(\lambda), u(\mu)) \\ = \; & d(H(u(\lambda), \lambda), H(u(\mu), \mu)) \\ \leq \; & d(H(u(\lambda), \lambda), H(u(\mu), \lambda)) + d(H(u(\mu), \lambda), H(u(\mu), \mu)) \\ \leq \; & M\, d(u(\lambda), u(\mu)) + d(H(u(\mu), \lambda), H(u(\mu), \mu)). \end{aligned}$$

By (a3) this implies

$$d(u(\lambda), u(\mu)) \leq (I - M)^{-1} d(H(u(\mu), \lambda), H(u(\mu), \mu)) \to 0$$

as $\lambda \to \mu$.

Step 4. Finding r. For any $\mu \in [0,1]$ and each $i \in \{1, 2, ..., n\}$ let

$$r_i(\mu) = \inf \{d_i(u, u(\mu)) : u \in X \setminus U\}.$$

Here $d = (d_1, d_2, ..., d_n)$. Since $u(\mu) \in U$ and U is open, $r_i(\mu) > 0$. We claim that

$$\inf \{r_i(\mu) : \mu \in [0,1]\} > 0. \tag{10.11}$$

To prove this let us assume the contrary. There are then $\mu_k \in [0,1]$ such that $r_i(\mu_k) \to 0$ as $k \to \infty$. Clearly we may assume that $\mu_k \to \mu$ for some $\mu \in [0,1]$. Then from the continuity of $u(\lambda)$ we have

$$d_i(u(\mu_k), u(\mu)) < \frac{1}{2} r_i(\mu), \quad k \geq k_1. \tag{10.12}$$

On the other hand, since $r_i(\mu_k) \to 0$ as $k \to \infty$,

$$r_i(\mu_k) < \frac{1}{2} r_i(\mu), \quad k \geq k_2. \tag{10.13}$$

Let $k_0 = \max\{k_1, k_2\}$. By (10.13) and the definition of $r_i(\mu_{k_0})$ as infimum, there is $u \in X \setminus U$ with

$$d_i(u, u(\mu_{k_0})) < \frac{1}{2} r_i(\mu). \tag{10.14}$$

Then, from (10.12) and (10.14), we obtain

$$\begin{aligned} d_i(u, u(\mu)) &\leq d_i(u, u(\mu_{k_0})) + d_i(u(\mu_{k_0}), u(\mu)) \\ &< 2 \cdot \frac{1}{2} r_i(\mu) = r_i(\mu), \end{aligned}$$

a contradiction. Thus (10.11) holds as claimed. Now we choose any $r_i > 0$ less than the infimum in (10.11), with the convention that $r_i = \infty$ if the infimum equals infinity; this happens when $U = X$ and so all the r_i's are finite or all equal infinity. Then take $r = (r_1, r_2, ..., r_n)$. If $r \in (0, \infty)^n$ we may assume that $r \in P_M$ (otherwise, replace it by the vector $\varepsilon (I - M)^{-1} r$ with $\varepsilon > 0$ small enough so that $\varepsilon (I - M)^{-1} r \leq r$).

Step 5. Finding of m and $0 < \lambda_1 < \lambda_2 < ..., \lambda_{m-1} < 1$. If $r = \infty$ take $h = 1$; if $r \in P_M$ let $h = \eta(r)$, where $\eta(r)$ is chosen as in (10.10). Then by what was shown at the end of Step 1 for each $\mu \in [0, 1]$,

$$\begin{cases} d(u, u(\mu)) \leq r \quad \text{and} \quad |\lambda - \mu| \leq h \quad \text{imply} \\ (H_\lambda^k(u))_{k\geq 0} \text{ is well defined and} \\ d(H_\lambda^k(u), u(\lambda)) \leq M^k (I - M)^{-1} d(u, H_\lambda(u)), \quad k \in \mathbf{N}. \end{cases} \tag{10.15}$$

Now we choose any partition

$$0 = \lambda_0 < \lambda_1 < ... < \lambda_{m-1} < \lambda_m = 1$$

of $[0, 1]$ such that $\lambda_{j+1} - \lambda_j \leq h$, $j = 0, 1, ..., m - 1$ (it is clear that $m = 1$ when $r = \infty$).

Step 6. Finding of the integers $k_1, k_2, ..., k_{m-1}$. From

$$d(u_{1,0}, u(0)) = d(u_0, u(0)) \leq r, \quad \lambda_1 - \lambda_0 \leq h,$$

by (10.15) we have that $(u_{1,i})_{i\geq 0}$ is well defined and satisfies (10.9). By (10.9), we may choose $k_1 \in \mathbf{N}$ such that

$$d(u_{1,k_1}, u(\lambda_1)) \leq r.$$

Now

$$d(u_{2,0}, u(\lambda_1)) = d(u_{1,k_1}, u(\lambda_1)) \leq r, \quad \lambda_2 - \lambda_1 \leq h,$$

and we repeat the above argument in order to show that $(u_{2,i})_{i\geq 0}$ is well defined and satisfies (10.9). In general, at step j $(1 \leq j \leq m - 1)$ we choose $k_j \in \mathbf{N}$ such that

$$d(u_{j,k_j}, u(\lambda_j)) \leq r.$$

Then since

$$d(u_{j+1,0}, u(\lambda_j)) = d(u_{j,k_j}, u(\lambda_j)) \leq r, \quad \lambda_{j+1} - \lambda_j \leq h,$$

by (10.15) we have that the sequence $(u_{j+1,i})_{i\geq 0}$ is well defined and satisfies (10.9). ■

The above proof yields the following algorithm for the approximation of $u(1)$ under the assumptions of Theorem 10.2:

Suppose we know r and h and we wish to obtain an approximation $\overline{u}_1$ of $u(1)$ with

$$d(\overline{u}_1, u(1)) \leq \varepsilon$$

for some $\varepsilon \in (0,\infty)$. Then we choose any partition

$$0 = \lambda_0 < \lambda_1 < \lambda_2 < ... < \lambda_{m-1} < \lambda_m = 1$$

of $[0,1]$ with $\lambda_{j+1} - \lambda_j \leq h$, $j = 0,1,...,m-1$, any element u_0 with $d(u_0, u(0)) \leq r$ and we follow the next:

Iterative procedure:

Set $k_0 := 0$ and $u_{0,k_0} := u_0$;
For $j := 1$ to $m-1$ do
 $u_{j,0} := u_{j-1,k_{j-1}}$
 $i := 0$
 While $M^i (I-A)^{-1} d(u_{j,0}, H_{\lambda_j}(u_{j,0})) \nleq r$
  $u_{j,i+1} := H_{\lambda_j}(u_{j,i})$
  $i := i+1$
 $k_j = i$
Set $i := 0$
While $M^i (I-M)^{-1} d(u_{m,0}, H_1(u_{m,0})) \nleq \varepsilon$
 $u_{m,i+1} = H_1(u_{m,i})$
 $i := i+1$
Finally take $\overline{u}_1 = u_{m,i}$.

Notice for $n = 1$, Theorem 10.2 reduces to Corollary 2.5 in Precup [50].

For other continuation results involving generalized contractions, see Agarwal–O'Regan [1], Frigon [23] and Precup [53]. For some related topics and applications, see Rus [56], Şerban [58] and Tsachev–Angelov [59].

10.3 Hammerstein Integral Equations with Matrix Kernels

In this section we give an application of Theorem 10.2 to the Hammerstein integral equation in $\mathbf{R}^n$

$$u(x) = \int_\Omega \kappa(x,y)\, f(y, u(y))\, dy, \quad x \in \Omega, \tag{10.16}$$

in the case that the kernel κ has matrix-values, i.e.,

$$\kappa : \Omega^2 \to \mathbf{M}_{n\times n}(\mathbf{R}), \quad \kappa = [\kappa_{ij}].$$

The usual Hammerstein equation in $\mathbf{R}^n$ with a scalar kernel appears as a particular case of (10.16) when $\kappa_{ij} = 0$ for $i \neq j$ and $\kappa_{ii} = \kappa_{jj}$ for all $i,j \in \{1,2,...,n\}$.

The simplest examples of problems which allow us to systems of the form (10.16) with matrix kernels are the boundary value problems for differential equations of order ≥ 2. For instance, the problem

$$\begin{cases} u'' = g(x, u, u'), & x \in [0,1] \\ u(0) = 0, \ u'(1) = 0 \end{cases}$$

can be put in the form (10.16) if we let $n = 2,\ u_1 = u,\ u_2 = u'$,

$$\kappa_{11}(x,y) = \begin{cases} 1, & y \leq x \\ 0, & y > x \end{cases}, \quad \kappa_{22}(x,y) = \begin{cases} 0, & y \leq x \\ -1, & y > x \end{cases}$$

$$\kappa_{12} = \kappa_{21} = 0,$$

and

$$f_1(x, u_1, u_2) = u_2, \quad f_2(x, u_1, u_2) = g(x, u_1, u_2).$$

Before we state the main result we introduce the following notations. For an element $z \in \mathbf{R}^n$, $z = (z_1, z_2, ..., z_n)$ we let

$$\|z\| = (|z_1|, |z_2|, ..., |z_n|).$$

Also, for a function $u \in L^p(\Omega; \mathbf{R}^n)$ $(\Omega \subset \mathbf{R}^N$ open, $1 \leq p \leq \infty)$, $u = (u_1, u_2, ..., u_n)$ we let

$$\|u\|_p = \left(|u_1|_p, |u_2|_p, ..., |u_n|_p\right).$$

Clearly $\|.\|$, $\|.\|_p$ are *vector-valued norms* in $\mathbf{R}^n$ and $L^p(\Omega; \mathbf{R}^n)$, respectively. Endowed with the vector-valued metric

$$d_p(u,v) = \|u - v\|_p,$$

$L^p(\Omega; \mathbf{R}^n)$ is a complete generalized metric space. Similarly, if $\Omega \subset \mathbf{R}^N$ is bounded open then $\left(C\left(\overline{\Omega}; \mathbf{R}^n\right), d_\infty\right)$ is a complete generalized metric space.

We now state and prove a general existence and uniqueness principle for (10.16).

Theorem 10.3 *Let $\Omega \subset \mathbf{R}^N$ be a bounded open set, $\kappa : \Omega^2 \to \mathbf{M}_{n\times n}(\mathbf{R})$ measurable and $f : \Omega \times \mathbf{R}^n \to \mathbf{R}^n$. Assume that there are $p \in [1, \infty]$,*

$q \in [1, \infty)$, $p \geq q$, and an open subset U of $\left(L^p(\Omega; \mathbf{R}^n), \|.\|_p\right)$ containing the origin, such that the following conditions are satisfied:

$$\begin{cases} (a) \text{ if } 1 \leq p < \infty \text{ then } \kappa_{ij}(x,.) \in L^r(\Omega) \text{ for a.e. } x \in \Omega, \text{ and} \\ \text{the map } x \longmapsto |\kappa_{ij}(x,.)|_r \text{ belongs to } L^p(\Omega) \ (1/q + 1/r = 1); \\ (b) \text{ if } p = \infty \text{ then } \kappa_{ij}(x,.) \in L^r(\Omega) \text{ for every } x \in \Omega, \text{ and} \\ \text{the map } x \longmapsto \kappa_{ij}(x,.) \text{ is continuous from } \overline{\Omega} \text{ to } L^r(\Omega); \end{cases} \tag{10.17}$$

$$\begin{cases} f \text{ satisfies the Carathéodory conditions, } f(.,0) \in L^q(\Omega; \mathbf{R}^n) \text{ and} \\ \|f(x,z_1) - f(x,z_2)\| \leq L(x) \|z_1 - z_2\| \text{ for a.e. } x \in \Omega, \\ \text{all } z_1, z_2 \in \mathbf{R}^n, \text{ and some } L \in L^{pq/(p-q)}(\Omega; \mathbf{M}_{n\times n}(\mathbf{R}_+)). \end{cases} \tag{10.18}$$

Let $M = [M_{ik}]$,

$$M_{ik} = \sum_{j=1}^{n} \left| |\kappa_{ij}(x,.)|_r \right|_p |L_{jk}|_{pq/(p-q)}$$

and assume M is convergent to zero. In addition assume that

$$u \in U \tag{10.19}$$

for any solution $u \in \overline{U}$ to

$$u(x) = \lambda \int_\Omega \kappa(x,y) f(y, u(y))\, dy \quad \text{a.e. on } \Omega, \tag{10.20}$$

for each $\lambda \in [0,1]$. Then (10.16) has a unique solution $u \in U \subset L^p(\Omega; \mathbf{R}^n)$. Moreover, $u \in C(\overline{\Omega}; \mathbf{R}^n)$ for $p = \infty$.

Proof. We note that by $pq/(p-q)$ we mean ∞ if $p = q$, and q if $p = \infty$. We apply Theorem 10.2 to $X = L^p(\Omega; \mathbf{R}^n)$ with vector-valued norm $\|.\|_p$ and $H : \overline{U} \times [0,1] \to L^p(\Omega; \mathbf{R}^n)$ given by

$$H(u,\lambda)(x) = \lambda \int_\Omega \kappa(x,y) f(y,u(y))\, dy \quad (x \in \Omega).$$

From (10.18) we have

$$\|f(x,z)\| \leq \|f(x,0)\| + L(x) \|z\|.$$

Hence

$$|f_i(x,z)| \leq |f_i(x,0)| + \sum_{j=1}^{n} L_{ij}(x) |z_j|. \tag{10.21}$$

By Young's inequality

$$L_{ij}(x)\,|z_j| \le \frac{p-q}{p} L_{ij}(x)^{p/(p-q)} + \frac{q}{p}\,|z_j|^{p/q}.$$

Since $f_i(.,0)$, $L_{ij}(.)^{p/(p-q)} \in L^q(\Omega)$, from (10.21) we obtain that

$$|f(x,z)| \le g(x) + c\,|z|^{p/q}$$

for some $g \in L^q(\Omega)$ and $c \ge 0$. Hence f is (p,q)-Carathéodory, and so the Nemytskii operator associated to f maps $L^p(\Omega;\mathbf{R}^n)$ into $L^q(\Omega;\mathbf{R}^n)$. From (10.17) we see that the Fredholm linear integral operators of kernels κ_{ij} maps $L^q(\Omega;\mathbf{R}^n)$ into $L^p(\Omega;\mathbf{R}^n)$. Hence H is well defined. Furthermore,

$$\begin{aligned}
&|H_i(u,\lambda)(x) - H_i(v,\lambda)(x)| \\
\le\; & \int_\Omega \sum_{j=1}^n |\kappa_{ij}(x,y)|\,|f_j(y,u(y)) - f_j(y,v(y))|\,dy \\
\le\; & \int_\Omega \sum_{j=1}^n |\kappa_{ij}(x,y)| \sum_{k=1}^n L_{jk}(y)\,|u_k(y) - v_k(y)|\,dy \\
\le\; & \sum_{k=1}^n \sum_{j=1}^n |\kappa_{ij}(x,.)|_r\,|L_{jk}|_{pq/(p-q)}\,|u_k - v_k|_p\,.
\end{aligned}$$

Consequently

$$\begin{aligned}
|H_i(u,\lambda) - H_i(v,\lambda)|_p &\le \sum_{k=1}^n \sum_{j=1}^n \left||\kappa_{ij}(x,.)|_r\right|_p |L_{jk}|_{pq/(p-q)}\,|u_k - v_k|_p \\
&= \sum_{k=1}^n M_{ik}\,|u_k - v_k|_p\,.
\end{aligned}$$

Thus

$$\|H(u,\lambda) - H(v,\lambda)\|_p \le M\,\|u - v\|_p\,.$$

Now the conclusion follows from Theorem 10.2. The cases $p = \infty$ and $p = q$ are left to the reader. ■

Notice that for Volterra–Hammerstein integral equations with matrix kernels one can use a similar argument to that in the proof of Theorem 5.6 in order to apply directly Perov's theorem.

For an extension of Theorem 10.2 to spaces with two vector-valued metrics and some applications to systems of abstract Hammerstein integral equations, see O'Regan–Precup [39].

Chapter 11

Monotone Iterative Methods

The basic notion in this chapter is that of an ordered Banach space. We try to localize solutions of an operator equation $u = T(u)$ in an ordered interval $[u_0, v_0]$ of an ordered Banach space X. In addition we look for solutions which are limits of increasing or decreasing sequences of elements of X. The basic property of the operator T is monotonicity. This combined with certain properties of the ordered Banach space X guarantees the convergence of monotone sequences. Thus we may say that this chapter explores the contribution of monotonicity to compactness.

11.1 Ordered Banach Spaces

There is a vast literature in ordered linear spaces and in ordered Banach spaces, in particular. We refer the reader to the books of Cristescu [17] and to Chapter 6 in Deimling [20]. Our aim for this section is just to introduce the reader to the theory of ordered Banach spaces and of monotone (in the sense of order) nonlinear operators. We shall restrict ourselves to those notions and results which will be used in the next sections.

Definition 11.1 Let X be a linear space. By a *cone* K of X we understand a convex subset of X such that $\lambda K \subset K$ for all $\lambda \in R_+$ and $K \cap (-K) = \{0\}$.

The proof of the following proposition is immediate and is left to the reader.

Proposition 11.1 *Let X be a linear space and $K \subset X$ be a cone. The* *relation* $\leq_K$ *on* X *defined by*

$$u \leq_K v \quad \textit{if and only if} \quad v - u \in K,$$

is an order (reflexive, antisymmetric, and transitive) relation on X *(called the order relation induced by* K*), compatible with the linear structure of* X*, i.e., whenever* $u_i,\ v_i \in X,\ u_i \leq_K v_i,\ i = 1,2,$ *and* $\lambda \in \mathbf{R}_+$*, we have*

$$u_1 + u_2 \leq_K v_1 + v_2, \quad \lambda u_1 \leq_K \lambda v_1.$$

Conversely, if $\leq$ *is an order relation on* X *compatible with the linear structure of* X*, then the set*

$$K_+ = \{u \in X : 0 \leq u\}$$

is a cone (called the positive cone) and $\leq = \leq_{K_+}$ *.*

The above proposition says that there is a one-to-one correspondence between the cones of X and the order relations on X which are compatible with the linear structure of X.

Throughout, we shall denote $\leq_K$ by $\leq$ for short. Also sometimes we shall write $u < v$ to indicate that $v - u \in \operatorname{int} K$.

Definition 11.2 A linear space endowed with a cone (equivalently, with an order relation compatible with its linear structure) is called an *ordered linear space.*

Let (X, K) be an ordered linear space. For any $u, v \in X$ we define the *order interval* $[u, v]$ by

$$\begin{aligned}[u,v] &= (u + K) \cap (v - K) \\ &= \{w \in X : u \leq w \leq v\}.\end{aligned}$$

Clearly $[u, v]$ is a convex set.

Definition 11.3 A Banach space endowed with a closed cone is called an *ordered Banach space.*

In an ordered Banach space any order interval $[u, v]$ is a closed convex subset. Moreover,

$$0 \leq u_k, \quad u_k \to u \ \text{ as } k \to \infty \ \text{ implies } \ 0 \leq u.$$

Consequently

$$u_k \le v_k,\ \ u_k \to u,\ \ v_k \to v \text{ implies } u \le v.$$

Thus in an ordered Banach space we may always pass to the limit in inequalities. All these statements follow immediately from the property of the cone of being closed. To have more we need additional properties of the cone.

Definition 11.4 Let (X, K) be an ordered Banach space and let $\le$ be the order relation induced by K. We say that:

1) K is a *normal cone* if

$$0 \le u_k \le v_k,\ \ v_k \to 0 \text{ implies } u_k \to 0.$$

2) K is a *regular cone* if every increasing sequence which is bounded from above is already convergent, i.e.

$$u_k \le u_{k+1} \le v \text{ for } k \in \mathbf{N} \text{ implies } (u_k) \text{ convergent.}$$

We have the following characterization theorem of normal cones.

Proposition 11.2 *Let (X, K) be an ordered Banach space with norm $|.|$. The following statements are equivalent:*

(a) K is a normal cone.

(b) The norm $|.|$ is semi-monotone, i.e., there exists a $\gamma > 0$ such that

$$0 \le u \le v \text{ implies } |u| \le \gamma\, |v|\,.$$

(c) $\inf\{|u + v| : \ u, v \in K,\ |u| = |v| = 1\} > 0.$

Proof. (a)⇒(b): Since K is a normal cone there is a number $\eta > 0$ such that

$$0 \le u \le v,\ \ |v| = \eta \text{ implies } |u| \le 1. \tag{11.1}$$

Indeed, otherwise we would find two sequences (u_k) and (v_k) with

$$0 \le u_k \le v_k,\ \ |v_k| = \frac{1}{k},\ \ |u_k| > 1.$$

But this is contrary to the definition of a normal cone. Let $\gamma = 1/\eta$. If $0 \le u \le v$ then

$$0 \le \eta\,|v|^{-1} u \le \eta\,|v|^{-1} v,\ \ \left|\eta\,|v|^{-1} v\right| = \eta.$$

By (11.1) this implies

$$\left|\eta\,|v|^{-1}\,u\right| \leq 1.$$

Hence $|u| \leq \eta^{-1}\,|v| = \gamma\,|v|$. Thus $|.|$ is semi-monotone.

(b)⇒(a): Immediate.

(b)⇒(c): For every $u, v \in K$ with $|u| = |v| = 1$, we have $0 \leq u \leq u+v$. Since $|.|$ is semi-monotone, this yields

$$1 = |u| \leq \gamma\,|u+v|.$$

Hence $|u+v| \geq 1/\gamma$. Thus

$$\inf\left\{|u+v| : u, v \in K,\ \ |u| = |v| = 1\right\} \geq \frac{1}{\gamma} > 0.$$

(c)⇒(b): Assume (b) is not true. Then there are two sequences (u_k), (v_k) with

$$0 \leq u_k \leq v_k, \quad |u_k| > k\,|v_k|.$$

Let

$$\overline{u}_k = |u_k|^{-1}\,u_k, \quad \overline{v}_k = |v_k|^{-1}\,v_k.$$

Also define

$$w_k = \left|k^{-1}\overline{v}_k - \overline{u}_k\right|^{-1}\left(k^{-1}\overline{v}_k - \overline{u}_k\right).$$

It is easy to check that

$$w_k \in K, \quad |w_k| = |\overline{u}_k| = 1, \quad |\overline{u}_k + w_k| \to 0.$$

This shows that (c) does not hold. ■

Proposition 11.3 *Any regular cone K is a normal cone.*

Proof. Assume the cone K is regular but not normal. Then statement (c) in Proposition 11.2 does not hold. Hence we may find two sequences (u_k), (v_k) such that

$$u_k, v_k \in K, \quad |u_k| = |v_k| = 1, \quad |u_k + v_k| \leq \frac{1}{2^k}.$$

Define the sequence (w_k) by

$$w_k = \sum_{j=1}^{k} u_j.$$

Clearly we have $w_k \in K$ and $w_k \leq w_{k+1}$. In addition

$$w_k \leq \sum_{j=1}^{\infty} (u_j + v_j) \in K.$$

Since the cone K is regular (w_k) is convergent. But

$$|w_{k+1} - w_k| = |u_{k+1}| = 1.$$

This contradiction proves that K is a normal cone. ■

Remark 11.1 In an ordered Banach space with normal cone, any order interval $[u, v]$ is a bounded set. Indeed, if $w \in [u, v]$ then $0 \leq w - u \leq v - u$. The norm being semi-monotone we have

$$|w - u| \leq \gamma |v - u|.$$

Example 11.1 Let $\Omega \subset R^N$ be bounded open. In $\left(C\left(\overline{\Omega}; \mathbf{R}^n\right), |.|_\infty\right)$, the set

$$K = \left\{u \in C\left(\overline{\Omega}; \mathbf{R}^n\right) : u(x) \geq 0 \text{ on } \overline{\Omega}\right\}$$

is a normal cone. However, it is not regular; for example, take $\Omega = (0, 1)$, $n = 1$, and

$$u_k(x) = \begin{cases} kx, & \text{for } x \in \left[0, \frac{1}{k}\right] \\ 1, & \text{for } x \in \left[\frac{1}{k}, 1\right]. \end{cases}$$

Clearly $u_k \leq u_{k+1} \leq 1$ but (u_k) does not converge in $C[0, 1]$

Example 11.2 Let $\Omega \subset R^N$ be open and let $1 \leq p < \infty$. The set

$$K = \{u \in L^p(\Omega; \mathbf{R}^n) : u(x) \geq 0 \text{ a.e. on } \Omega\}$$

is a regular cone of $L^p(\Omega; \mathbf{R}^n)$. For the proof use Beppo Levi's theorem (see Brezis [7], p 55).

11.2 Fixed Point Theorems for Monotone Operators

Definition 11.5 Let X be a set endowed with an order relation $\leq$, and let $T : D \subset X \to X$ be a map. We say that T is *increasing* (or *isotone*) if $T(u) \leq T(v)$ for $u, v \in D$ with $u \leq v$. The map T is said to be *decreasing* (or *anti-isotone*) if $T(v) \leq T(u)$ for $u, v \in D$ with $u \leq v$.

We now state and prove the *monotone iterative principle* for increasing operators in ordered Banach spaces.

Theorem 11.1 *Let (X, K) be an ordered Banach space, $[u_0, v_0] \subset X$ an order interval and $T : [u_0, v_0] \to [u_0, v_0]$ an increasing continuous operator. Assume one of the following conditions:*

(i) K is a regular cone.

(ii) K is a normal cone and T is completely continuous.

Then there exist $u^*, v^* \in [u_0, v_0]$ such that $u^* \leq v^*$, $T(u^*) = u^*$, $T(v^*) = v^*$, $(T^k(u_0))$ is increasing and convergent to u^*, and $(T^k(v_0))$ is decreasing and convergent to v^*. Moreover, any fixed point of T lies between u^* and v^*.

Proof. Notice the increasing operator T maps $[u_0, v_0]$ into itself if and only if

$$u_0 \leq T(u_0), \quad T(v_0) \leq v_0.$$

From

$$u_0 \leq T(u_0) \leq T(v_0) \leq v_0$$

we find

$$u_0 \leq T(u_0) \leq T^2(u_0) \leq T^2(v_0) \leq T(v_0) \leq v_0,$$

and finally

$$\begin{aligned} u_0 &\leq T(u_0) \qquad (11.2) \\ &\vdots \\ &\leq T^k(u_0) \leq T^{k+1}(u_0) \\ &\vdots \\ &\leq T^{k+1}(v_0) \leq T^k(v_0) \\ &\vdots \\ &\leq T(v_0) \leq v_0. \end{aligned}$$

Assume (i). Then (11.2) implies the convergence of the sequences $(T^k(u_0))$ and $(T^k(v_0))$. Let u^*, v^* be their limits. Then from

$$T^{k+1}(u_0) = T\left(T^k(u_0)\right)$$

we deduce that $u^* = T(u^*)$ since T is assumed to be continuous. Similarly, $v^* = T(v^*)$.

Assume (ii). The order interval $[u_0, v_0]$ being bounded (see Remark 11.1) and T being completely continuous, we have that $T([u_0, v_0])$ is relatively compact. Consequently $(T^k(u_0))_{k\geq 1}$ has a convergent subsequence $(T^{k_j}(u_0))_{j\geq 1}$. Let u^* be its limit. Clearly $u^* \in [u_0, v_0]$. We claim that the whole sequence $(T^k(u_0))_{k\geq 1}$ converges to u^*. Indeed, for each $\varepsilon > 0$ there is a j_ε such that

$$\left|T^{k_{j_\varepsilon}}(u_0) - u^*\right| \leq \frac{\varepsilon}{1+\gamma}.$$

Here the constant $\gamma > 0$ comes from the semi-monotonicity of the norm. On the other hand, for $i \geq k_{j_\varepsilon}$ we have

$$T^{k_{j_\varepsilon}}(u_0) \leq T^i(u_0) \leq u^*,$$

that is

$$0 \leq T^i(u_0) - T^{k_{j_\varepsilon}}(u_0) \leq u^* - T^{k_{j_\varepsilon}}(u_0).$$

Then

$$\left|T^i(u_0) - T^{k_{j_\varepsilon}}(u_0)\right| \leq \gamma\left|u^* - T^{k_{j_\varepsilon}}(u_0)\right|.$$

Hence

$$\begin{aligned} \left|T^i(u_0) - u^*\right| &\leq \left|T^i(u_0) - T^{k_{j_\varepsilon}}(u_0)\right| + \left|T^{k_{j_\varepsilon}}(u_0) - u^*\right| \\ &\leq (\gamma+1)\left|T^{k_{j_\varepsilon}}(u_0) - u^*\right| \\ &\leq \varepsilon. \end{aligned}$$

Thus $T^i(u_0) \to u^*$ as claimed. Use similar arguments to prove the convergence of $(T^k(v_0))$.

Now assume that $w \in [u_0, v_0]$ is a fixed point of T. Then

$$u_0 \leq T(u_0) \leq T(w) = w \leq T(v_0) \leq v_0$$

and in general

$$T^k(u_0) \leq w \leq T^k(v_0).$$

Letting $k \to \infty$ we obtain $u^* \leq w \leq v^*$. ■

We shall use the following terminology:

1) An element u_0 for which $u_0 \leq T(u_0)$ is said to be a *lower solution* of the operator equation $u = T(u)$.

2) An element v_0 satisfying $T(v_0) \leq v_0$ is said to be an *upper solution* of the equation $u = T(u)$.

3) The fixed point u^* given by Theorem 11.1 is the *minimal solution* of the equation $u = T(u)$ in $[u_0, v_0]$, whilst v^* is the *maximal solution* in $[u_0, v_0]$.

Remark 11.2 Under the assumptions of Theorem 11.1, if $u^* = v^*$ then

$$T^k(w) \to u^* \quad \text{for all} \quad w \in [u_0, v_0].$$

The next result is an existence and uniqueness theorem.

Theorem 11.2 *Let (X, K) be an ordered Banach space and let $[u_0, v_0] \subset X$ be an order interval with $0 \le u_0 \le v_0$ such that there is a number λ with*

$$\lambda > 0, \quad \lambda v_0 \le u_0. \tag{11.3}$$

Assume $T : [0, v_0] \to X$ is increasing on $[0, v_0]$, continuous on $[u_0, v_0]$, $T([u_0, v_0]) \subset [u_0, v_0]$ and that there is a function $\phi : (0,1) \to \mathbf{R}$ such that

$$\phi(\lambda) > \lambda, \quad \phi(\lambda) T(u) \le T(\lambda u) \tag{11.4}$$

for all $\lambda \in (0,1)$, $u \in [u_0, v_0]$. In addition assume one of the following conditions:

(i) K is a regular cone.

(ii) K is a normal cone and T is completely continuous.

Then T has a unique fixed point u^* in $[u_0, v_0]$, $(T^k(u_0))$ is increasing and convergent to u^*, $(T^k(v_0))$ is decreasing and convergent to u^*, and

$$T^k(w) \to u^* \quad \textit{for all} \quad w \in [u_0, v_0].$$

Proof. Let u^*, v^* be the fixed point of T given by Theorem 11.1. By (11.3),

$$\lambda v^* \le \lambda v_0 \le u_0 \le u^*.$$

Hence the set $\{\lambda \in (0,1] : \lambda v^* \le u^*\}$ is nonempty. Let

$$\lambda^* = \sup\{\lambda \in (0,1] : \lambda v^* \le u^*\}.$$

Clearly, $\lambda^* v^* \le u^*$. If $\lambda^* = 1$ we have finished. Assume $\lambda^* < 1$. Then using (11.4) and the monotonicity of T on $[0, v_0]$, we obtain

$$\begin{aligned} \phi(\lambda^*) v^* &= \phi(\lambda^*) T(v^*) \\ &\le T(\lambda^* v^*) \le T(u^*) = u^*. \end{aligned}$$

Since $\phi(\lambda^*) > \lambda^*$, this is in contradiction with the definition of λ^*. ■

A similar existence and uniqueness result holds for decreasing operators.

Theorem 11.3 *Let* (X, K) *be an ordered Banach space and let* $[u_0, v_0] \subset X$ *be an order interval with* $0 \leq u_0 \leq v_0$ *such that there is a number* λ *with*

$$\lambda > 0, \quad \lambda v_0 \leq u_0. \tag{11.5}$$

Assume $T : [0, v_0] \to X$ *is decreasing on* $[0, v_0]$, *continuous on* $[u_0, v_0]$, $T([u_0, v_0]) \subset [u_0, v_0]$ *and that there is a function* $\chi : (0,1) \to (0, \infty)$ *such that*

$$\chi(\lambda) < \frac{1}{\lambda}, \quad T(\lambda u) \leq \chi(\lambda) T(u) \tag{11.6}$$

for all $\lambda \in (0,1)$, $u \in [u_0, v_0]$. *In addition assume one of the following conditions:*

(*i*) K *is a regular cone.*

(*ii*) K *is a normal cone and* T *is completely continuous.*

Then T *has a unique fixed point* u^* *in* $[u_0, v_0]$. *Moreover, the sequences* $(T^{2k}(u_0))$ *and* $(T^{2k+1}(v_0))$ *are increasing and convergent to* u^*, *whilst* $(T^{2k+1}(u_0))$ *and* $(T^{2k}(v_0))$ *are decreasing and convergent to* u^*. *Also*

$$T^k(w) \to u^* \quad \textit{for all} \quad w \in [u_0, v_0].$$

Proof. Since T is decreasing and $T([u_0, v_0]) \subset [u_0, v_0]$, we have

$$u_0 \leq T(v_0) \leq T(u_0) \leq v_0.$$

This implies

$$u_0 \leq T(v_0) \leq T^2(u_0) \leq T^2(v_0) \leq T(u_0) \leq v_0.$$

Finally

$$\begin{aligned} u_0 &\leq T(v_0) \\ &\vdots \\ &\leq T^{2k}(u_0) \leq T^{2k+1}(v_0) \\ &\vdots \\ &\leq T^{2k+1}(u_0) \leq T^{2k}(v_0) \\ &\vdots \\ &\leq T(u_0) \leq v_0. \end{aligned}$$

As in the proof of Theorem 11.1,

$$T^{2k}(u_0) \to u^*, \quad T^{2k+1}(u_0) \to v^*, \quad u_0 \leq u^* \leq v^* \leq v_0.$$

We have

$$T(u^*) = v^*, \quad T(v^*) = u^*.$$

Let

$$\lambda^* = \sup\{\lambda \in (0,1] : \lambda v^* \le u^*\}.$$

Then $\lambda^* v^* \le u^*$. We show that $\lambda^* = 1$. Assume $\lambda^* < 1$. Then using (11.6) we deduce that

$$\begin{aligned} v^* &= T(u^*) \\ &\le T(\lambda^* v^*) \le \chi(\lambda^*) T(v^*) \\ &= \chi(\lambda^*) u^*. \end{aligned}$$

Since $\chi(\lambda^*) < 1/\lambda^*$, this contradicts the definition of λ^*. Hence $\lambda^* = 1$ and so $u^* = v^*$. Now, from

$$T^{2k}(u_0) \le T^{2k+1}(v_0) \le T^{2k}(v_0) \le T^{2k-1}(u_0)$$

letting $k \to \infty$ we obtain

$$T^{2k+1}(v_0) \to u^*, \quad T^{2k}(v_0) \to u^*.$$

Finally, it is easy to conclude that $T^k(w) \to u^*$ for every $w \in [u_0, v_0]$. ■

The results presented in this section are more or less in connection with those established by Amman [2], Guo–Lakshmikantham [27], Heikkilä–Lakshmikantham [30] and Krasnoselskii [33].

11.3 Monotone Iterative Technique for Fredholm Integral Equations

Consider the equation

$$u(x) = \int_\Omega h(x, y, u(y))\, dy, \quad x \in \Omega \tag{11.7}$$

where $\Omega \subset \mathbf{R}^N$ is bounded open and $h : \overline{\Omega}^2 \times \mathbf{R}^n \to \mathbf{R}^n$.

Theorem 11.4 *Let $u_0, v_0 \in C(\overline{\Omega}; \mathbf{R}^n)$ be such that $u_0(x) \le v_0(x)$ and*

$$u_0(x) \le \int_\Omega h(x, y, u_0(y))\, dy, \quad \int_\Omega h(x, y, v_0(y))\, dy \le v_0(x)$$

for all $x \in \overline{\Omega}$. *Assume* $h \in C\left(\overline{\Omega}^2 \times \mathbf{R}^n; \mathbf{R}^n\right)$ *and*

$$h(x, y, z) \leq h(x, y, z') \tag{11.8}$$

for all $z, z' \in \mathbf{R}^n$ *satisfying* $u_0(x) \leq z \leq z' \leq v_0(x)$, *and all* $x, y \in \overline{\Omega}$. *Then the sequence* $(u_k)_{k \geq 0}$ *given by*

$$u_{k+1}(x) = \int_\Omega h(x, y, u_k(y))\, dy, \quad k \in \mathbf{N}$$

is increasing and converges to a solution $u^* \in C\left(\overline{\Omega}; \mathbf{R}^n\right)$ *of* (11.7). *Also the sequence* $(v_k)_{k \geq 0}$ *defined as*

$$v_{k+1}(x) = \int_\Omega h(x, y, v_k(y))\, dy, \quad k \in \mathbf{N}$$

is decreasing and converges to a solution $v^* \in C\left(\overline{\Omega}; \mathbf{R}^n\right)$ *of* (11.7). *In addition,* $u_0 \leq u^* \leq v^* \leq v_0$ *and any solution* $w \in C\left(\overline{\Omega}; \mathbf{R}^n\right)$ *with* $u_0 \leq w \leq v_0$ *satisfies* $u^* \leq w \leq v^*$.

Proof. Apply Theorem 11.1 to $X = C\left(\overline{\Omega}; \mathbf{R}^n\right)$ endowed with the positive cone K (see Example 11.1) and to the operator $T : C\left(\overline{\Omega}; \mathbf{R}^n\right) \to C\left(\overline{\Omega}; \mathbf{R}^n\right)$ given by

$$T(u)(x) = \int_\Omega h(x, y, u(y))\, dy.$$

Here K is a normal cone and T is completely continuous. ∎

For a function $u : D \to \mathbf{R}^n$, notation $u > 0$ stands for $u(x) > 0$ on D, i.e., $u_i(x) > 0$ on D for $i = 1, 2, ..., n$.

Theorem 11.5 *Assume all the assumptions of Theorem* 11.4 *are satisfied. In addition assume that* $u_0 > 0$, (11.8) *holds for all* $z, z' \in \mathbf{R}^n$ *satisfying* $0 \leq z \leq z' \leq v_0(x)$ *and all* $x, y \in \overline{\Omega}$, *and there is a function* $\phi : (0,1) \to \mathbf{R}$ *such that* $\phi(\lambda) > \lambda$ *and*

$$\phi(\lambda)\, h(x, y, z) \leq h(x, y, \lambda z) \tag{11.9}$$

for $\lambda \in (0,1)$, $z \in \mathbf{R}^n$ *with* $u_0(x) \leq z \leq v_0(x)$, *and all* $x, y \in \overline{\Omega}$. *Then* (11.7) *has a unique solution* u^* *in* $[u_0, v_0]$ *and*

$$T^k(w) \to u^* \quad \text{for all} \quad w \in [u_0, v_0],$$

where $\left(T^k(u_0)\right)$ *is increasing and* $\left(T^k(v_0)\right)$ *is decreasing.*

Proof. Apply Theorem 11.2. Condition (11.3) holds for

$$\lambda = \min\left\{\frac{u_{0i}(x)}{v_{0i}(x)} : x \in \overline{\Omega},\ i \in \{1, 2, ..., n\}\right\},$$

where $u_0 = (u_{01}, u_{02}, ..., u_{0n})$ and $v_0 = (v_{01}, v_{02}, ..., v_{0n})$. ■

A better result can be established when u_0 and v_0 are constants. For a real number a, we shall also denote by a the vector $(a, a, ..., a) \in \mathbf{R}^n$ and the corresponding constant vector-valued function. Also the notation $[a, b]$ stands for the order interval in $\mathbf{R}^n$ and in any space of functions with values in $\mathbf{R}^n$, ordered by the usual positive cone.

Theorem 11.6 *Let $a, b \in \mathbf{R}$, $0 < a < b$ and*

$$a \leq \int_\Omega h(x, y, a)\, dy, \quad \int_\Omega h(x, y, b)\, dy \leq b$$

for all $x \in \overline{\Omega}$. Assume $h \in C\left(\overline{\Omega}^2 \times [a, b]; \mathbf{R}^n\right)$ and

$$h(x, y, z) \leq h(x, y, z') \tag{11.10}$$

for all $z, z' \in \mathbf{R}^n$ satisfying $a \leq z \leq z' \leq b$ and all $x, y \in \overline{\Omega}$. In addition assume that there is a function $\phi : [a/b, 1) \to \mathbf{R}$ such that $\phi(\lambda) > \lambda$ and

$$\phi(\lambda)\, h(x, y, z) \leq h(x, y, \lambda z) \tag{11.11}$$

for $\lambda \in [a/b, 1)$, $z \in [a, b]$ with $\lambda z \geq a$ and all $x, y \in \overline{\Omega}$. Then (11.7) has a unique solution u^* in $[a, b]$,

$$T^k(w) \to u^* \quad \textit{for all} \quad w \in [a, b],$$

where $\left(T^k(a)\right)$ is increasing and $\left(T^k(b)\right)$ is decreasing.

Proof. Let u^*, v^* be the minimal and maximal solutions of (11.7) in $[a, b]$ guaranteed by Theorem 11.1. Let

$$\lambda = \min\left\{\frac{u_i^*(x)}{v_i^*(x)} : x \in \overline{\Omega},\ i \in \{1, 2, ..., n\}\right\}.$$

Clearly, $\lambda \in [a/b, 1]$. If $\lambda = 1$, then $u^* = v^*$ and we are finished. Assume $\lambda < 1$. Define $u = (u_1, u_2, ..., u_n)$ by

$$u_i(x) = \max\{a, \lambda v_i^*(x)\} = \lambda \max\left\{\frac{a}{\lambda}, v_i^*(x)\right\}.$$

Let $v = (v_1, v_2, ..., v_n)$ be defined by

$$v_i(x) = \max\left\{\frac{a}{\lambda}, v_i^*(x)\right\}.$$

We have $a \le u_i(x) \le u_i^*(x)$. Hence $u \in [a, u^*]$. Also $v \in [v^*, b]$. Using the monotonicity of T on $[a, b]$ and (11.11), we obtain

$$\begin{aligned}\phi(\lambda) v^* &= \phi(\lambda) T(v^*) \\ &\le \phi(\lambda) T(v) \le T(\lambda v) = T(u) \\ &\le T(u^*) = u^*.\end{aligned}$$

This is in contradiction with the definition of λ since $\phi(\lambda) > \lambda$. ∎

Example 11.3 The equation

$$u(x) = \int_\Omega \kappa(x, y) u(y)^\alpha dy, \quad x \in \overline{\Omega} \tag{11.12}$$

has a unique solution $u \in C(\overline{\Omega})$ satisfying $a \le u(x) \le b$ on $\overline{\Omega}$, where $0 < a < b$, if $\kappa \in C\left(\overline{\Omega}^2; \mathbf{R}_+\right)$, $\alpha \in (0, 1)$ and

$$a^{1-\alpha} \le \int_\Omega \kappa(x, y) dy \le b^{1-\alpha}, \quad x \in \overline{\Omega}.$$

Here $\phi(\lambda) = \lambda^\alpha$.

Similar results can be stated for the case when T is decreasing. We leave their proofs to the reader.

Theorem 11.7 *Let $u_0, v_0 \in C(\overline{\Omega}; \mathbf{R}^n)$ be such that $0 < u_0 \le v_0$ and*

$$u_0(x) \le \int_\Omega h(x, y, v_0(y)) dy, \quad \int_\Omega h(x, y, u_0(y)) dy \le v_0(x)$$

for all $x \in \overline{\Omega}$. Assume $h \in C\left(\overline{\Omega}^2 \times \mathbf{R}^n; \mathbf{R}^n\right)$ and

$$h(x, y, z') \le h(x, y, z) \tag{11.13}$$

for all $z, z' \in \mathbf{R}^n$ satisfying $0 \le z \le z' \le v_0(x)$, and all $x, y \in \overline{\Omega}$. In addition assume that there is a function $\chi : (0, 1) \to (0, \infty)$ such that $\chi(\lambda) < 1/\lambda$ and

$$h(x, y, \lambda z) \le \chi(\lambda) h(x, y, z) \tag{11.14}$$

for $\lambda \in (0,1)$, $z \in \mathbf{R}^n$ *with* $u_0(x) \le z \le v_0(x)$, *and all* $x, y \in \overline{\Omega}$. *Then* (11.7) *has a unique solution* u^* *in* $[u_0, v_0]$ *and*

$$T^k(w) \to u^* \quad \textit{for all} \quad w \in [u_0, v_0].$$

Moreover, the sequences $(T^{2k}(u_0))$, $(T^{2k+1}(v_0))$ *are increasing and the sequences* $(T^{2k+1}(u_0))$, $(T^{2k}(v_0))$ *are decreasing.*

Theorem 11.8 *Let* $a, b \in \mathbf{R}$, $0 < a < b$ *and*

$$a \le \int_\Omega h(x,y,b)\,dy, \quad \int_\Omega h(x,y,a)\,dy \le b$$

for all $x \in \overline{\Omega}$. *Assume* $h \in C\left(\overline{\Omega}^2 \times [a,b]; \mathbf{R}^n\right)$ *and*

$$h(x,y,z') \le h(x,y,z) \tag{11.15}$$

for all $z, z' \in \mathbf{R}^n$ *satisfying* $a \le z \le z' \le b$ *and all* $x, y \in \overline{\Omega}$. *In addition assume that there is a function* $\chi : [a/b, 1) \to (0, \infty)$ *such that* $\chi(\lambda) < 1/\lambda$ *and*

$$h(x,y,\lambda z) \le \chi(\lambda)\, h(x,y,z) \tag{11.16}$$

for $\lambda \in [a/b, 1)$, $z \in [a,b]$ *with* $\lambda z \ge a$, *and* $x, y \in \overline{\Omega}$. *Then* (11.7) *has a unique solution* u^* *in* $[a,b]$,

$$T^k(w) \to u^* \quad \textit{for all} \quad w \in [a,b],$$

the sequences $(T^{2k}(u_0))$, $(T^{2k+1}(v_0))$ *are increasing, and the sequences* $(T^{2k+1}(u_0))$, $(T^{2k}(v_0))$ *are decreasing.*

Example 11.4 The equation (11.12) has a unique solution $u \in C(\overline{\Omega})$ satisfying $a \le u(x) \le b$ on $\overline{\Omega}$, where $0 < a < b$, if $\kappa \in C\left(\overline{\Omega}^2; \mathbf{R}_+\right)$, $\alpha \in (-1, 0)$ and

$$ab^{-\alpha} \le \int_\Omega \kappa(x,y)\,dy \le a^{-\alpha} b, \quad x \in \overline{\Omega}.$$

Here $\chi(\lambda) = \lambda^\alpha$.

11.4 Minimal and Maximal Solutions of a Delay Integral Equation

In this section we present several existence, uniqueness, and approximation results for the initial valued problem and for the periodic problem concerning the delay integral equation in $\mathbf{R}^n$

$$u(t) = \int_{t-\tau}^{t} f(s, u(s))\, ds. \tag{11.17}$$

1. The Initial Value Problem

Here we are interested in continuous solutions u of (11.17), with $u(t) \geq a$ on $[0, t_1]$, such that

$$u(t) = \varphi(t), \quad t \in [-\tau, 0]. \tag{11.18}$$

We assume that $\varphi \in C([-\tau, 0]; \mathbf{R}^n)$ and

$$\varphi(0) = \int_{-\tau}^{0} f(s, \varphi(s))\, ds. \tag{11.19}$$

The assumptions are as follows:

(a1) $f \in C([-\tau, t_1] \times [a, \infty)^n; \mathbf{R}^n_+)$;
(a2) $\varphi \in C([-\tau, 0]; \mathbf{R}^n)$, satisfies (11.19) and $\varphi(t) \geq a$ on $[-\tau, 0]$;
(a3) there exists a function $g \in C([-\tau, t_1]; \mathbf{R}^n)$ such that

$$f(t, z) \geq g(t)$$

for all $t \in [-\tau, t_1]$, $z \in [a, \infty)^n$, and

$$\int_{t-\tau}^{t} g(s)\, ds \geq a$$

for all $t \in [0, t_1]$;

(a4) there exists a continuous nondecreasing function $\psi : (a, \infty) \to (0, \infty)$ such that

$$|f(t, z)| \leq \psi(|z|)$$

for all $t \in [0, t_1]$, $z \in [a, \infty)^n$, and

$$t_1 < \int_{b}^{\infty} \frac{1}{\psi(\sigma)} d\sigma,$$

where

$$b = \int_{-\tau}^{0} |f(s, \varphi(s))| \, ds.$$

Let $R_0 \in \mathbf{R}_+$ be such that

$$t_1 = \int_{b}^{R_0} \frac{1}{\psi(\sigma)} d\sigma. \tag{11.20}$$

Let $X = C([0, t_1]; \mathbf{R}^n)$ be endowed with the usual positive cone, let

$$K = \{u \in X : u \geq a\}$$

and let $T : K \to C([0, t_1]; \mathbf{R}^n)$ be given by

$$T(u)(t) = \int_{t-\tau}^{t} f(s, \widetilde{u}(s)) \, ds, \tag{11.21}$$

where $\widetilde{u}(t) = \varphi(t)$ for $t \in [-\tau, 0]$, $\widetilde{u}(t) = u(t)$ for $t \in [0, t_1]$.

Theorem 11.9 *Assume* (a1)–(a4) *are satisfied. In addition assume that* $f(t, z)$ *is increasing in* z *on* $[a, R_0]$. *Then* $T^k(a) \to u^*$ *uniformly on* $[0, t_1]$, u^* *is the minimal solution of* (11.17)–(11.18) *in* K, $|u^*(t)| \leq R_0$ *on* $[0, t_1]$, *and*

$$a \leq T(a) \leq \ldots \leq T^k(a) \leq \ldots \leq u^*.$$

Proof. In this situation a is a lower solution of the equation $u = T(u)$. By Theorem 4.6, (11.17)–(11.18) has at least one solution $v \in K$. In addition any solution v satisfies

$$|v(t)| \leq R_0 \quad \text{on} \quad [0, t_1].$$

Consequently, $v \leq R_0$. Now we apply Theorem 11.1 by setting $u_0 = a$ and $v_0 = v$, where $v \in K$ is any solution of the initial value problem. ■

The next result deals with the existence and approximation of the maximal solution in K of (11.17)–(11.18).

Theorem 11.10 *Assume* (a1)–(a4) *are satisfied. In addition assume that there is* $R \geq R_0$ *such that*

$$f\left(t, \widetilde{R}\right) \leq \tau^{-1} R, \quad t \in [-\tau, t_1] \tag{11.22}$$

(*i.e.*, $f(t, \varphi(t)) \leq \tau^{-1} R$ *for* $t \in [-\tau, 0]$ *and* $f(t, R) \leq \tau^{-1} R$ *for* $t \in [0, t_1]$), *and* $f(t, z)$ *is increasing in* z *on* $[a, R]$. *Then* $T^k(R) \to v^*$ *uniformly on* $[0, t_1]$, v^* *is the maximal solution in* K *of* (11.17)–(11.18), *and*

$$a \leq v^* \leq \ldots \leq T^{k+1}(R) \leq T^k(R) \leq \ldots \leq T(R) \leq R.$$

Proof. By (11.22), we have $T(R) \leq R$. Further the proof is analog with that of Theorem 11.1. Here we choose as upper solution the constant function $v_0 = R$ and as lower solution u_0, any solution in K to our problem. ∎

Theorem 11.11 *Assume the assumptions of Theorem 11.9 are satisfied. In addition assume $a > 0$ and that there is a function $\phi : [a/R_0, 1) \to \mathbf{R}$ such that for all $\lambda \in [a/R_0, 1)$, $t \in [0, t_1]$ and $z \in \mathbf{R}^n$ with $z \in [a, R_0]$ and $\lambda z \geq a$, one has*

$$\phi(\lambda) > \lambda, \quad \phi(\lambda) f(t, z) \leq f(t, \lambda z). \tag{11.23}$$

Then (11.17)–(11.18) *has a unique solution in K.*

Proof. Let u^* be the solution given by Theorem 11.9 and let $u \in K$ be any solution of (11.17)–(11.18). We will show that $u = u^*$. Let

$$\lambda = \min\left\{\frac{u_i^*(t)}{u_i(t)} : t \in [0, t_1],\ i \in \{1, 2, ..., n\}\right\}.$$

Since $a \leq u^* \leq u \leq R_0$, we have $\lambda \in [a/R_0, 1]$. As in the proof of Theorem 11.2, we deduce that $\lambda = 1$. Hence $u = u^*$. ∎

For $\phi(\lambda) = \lambda^\alpha$, $\alpha \in (0, 1)$, and $n = 1$, Theorem 11.11 becomes Theorem 4 in Precup [47]. Another example of function ϕ satisfying (11.23), is

$$\phi(\lambda) = \frac{\log(1 + a\lambda)}{\log(1 + a)}$$

for $f(t, z)$ of the form $q(t) \log(1 + |z|)$ (see Dads–Ezzinbi–Arino [18]).

Corollary 11.1 *Assume that all the assumptions of Theorems 11.10 and 11.11 are satisfied. Then* (11.17)–(11.18) *has a unique solution u^* in K, and for any $w \in C([0, t_1]; \mathbf{R}^n)$ with $w \in [a, R]$ one has $T^k(w) \to u^*$ uniformly on $[0, t_1]$.*

The next result refers to functions $f(t, z)$ which are decreasing in z.

Theorem 11.12 *Assume* (a1)–(a4) *are satisfied. Denote*

$$R = \max\left\{R_0,\ \max_{t \in [0, t_1]} |T(a)(t)|\right\}$$

and assume $f(t,z)$ *is decreasing in* z *for* $0 < a \leq z \leq R$. *Also assume that there is a function* $\chi : [a/R, 1) \to (0, \infty)$ *such that for all* $\lambda \in [a/R, 1)$, $t \in [0, t_1]$ *and* $z \in \mathbf{R}^n$ *with* $z \in [a, R]$ *and* $\lambda z \geq a$, *one has*

$$\chi(\lambda) < \frac{1}{\lambda}, \quad f(t, \lambda z) \leq \chi(\lambda) f(t, z). \tag{11.24}$$

Then (11.17)-(11.18) *has a unique solution* u^* *in* K, $T^k(w) \to u^*$ *uniformly on* $[0, t_1]$ *for every* $w \in [a, R]$, *the sequences* $(T^{2k}(a))$, $(T^{2k+1}(R))$ *are increasing, and the sequences* $(T^{2k+1}(a))$, $(T^{2k}(R))$ *are decreasing.*

Proof. Observe that $T([a, R]) \subset [a, R]$. ■

For $\chi(\lambda) = \lambda^\alpha$, $\alpha \in (-1, 0)$, and $n = 1$, Theorem 11.12 becomes Theorem 6 in Precup [47].

2. Periodic Solutions

We now present similar results for the periodic solutions of the equation (11.17).

We are interested in periodic continuous solutions u such that

$$a \leq u(t) \leq R \quad \text{for all } t \in \mathbf{R}.$$

The hypotheses are those from Section 4.5:

(h1) $f \in C(\mathbf{R} \times [a, \infty)^n; \mathbf{R}^n_+)$.

(h2) There is $\omega > 0$ such that $f(t + \omega, z) = f(t, z)$, $t \in \mathbf{R}$, $z \in [a, \infty)^n$.

(h3) There exists an ω-periodic function $g \in C(\mathbf{R}; \mathbf{R}^n)$ such that

$$f(t, z) \geq g(t)$$

for all $t \in \mathbf{R}$, $z \in [a, \infty)^n$, and

$$\int_{t-\tau}^{t} g(s)\, ds \geq a$$

for all $t \in \mathbf{R}$.

(h4) there is a number b with $a\sqrt{n} < b < R$, and a function $\psi \in C([a, R]; \mathbf{R}_+)$ with $\psi(t) > 0$ on $[b, R]$, such that

$$|f(t, z)| \leq \psi(|z|)$$

for all $t \in \mathbf{R}$ and $z \in [a, \infty)^n$ with $|z| \leq R$,

$$\omega \leq \int_b^R \frac{1}{\psi(\sigma)} d\sigma \tag{11.25}$$

and

$$|f(t,z)| \leq \frac{b}{\tau} \tag{11.26}$$

for all $t \in \mathbf{R}$ and $z \in [a,\infty)^n$ with $b \leq |z| \leq R$.

Let X be the Banach space of all continuous ω-periodic functions $u : \mathbf{R} \to \mathbf{R}^n$, ordered by the usual positive cone, and let

$$K = \{u \in X : u(t) \geq a \text{ on } [0,\omega]\}.$$

Under assumptions (h1)–(h4), the operator $T : [a,R] \to K$, given by

$$T(u)(t) = \int_{t-\tau}^{t} f(s,u(s))\,ds$$

is well defined and completely continuous.

Theorem 11.13 *Assume (h1)–(h4) are satisfied. In addition assume that* $a > 0$, $f(t,z)$ *is decreasing in* z *on* $[a,R]$, *and that there exists a function* $\chi : [a/R,1) \to \mathbf{R}$ *satisfying* (11.24) *for all* $t \in [0,\omega]$, $\lambda \in [a/R,1)$ *and* $z \in \mathbf{R}^n$ *with* $z \in [a,R]$ *and* $\lambda z \geq a$. *If*

$$T^2(R) \leq R, \tag{11.27}$$

then (11.17) *has a unique* ω*-periodic solution* $u^* \in [a,R]$. *Moreover,*

$$T^k(R) \to u^*$$

uniformly on $[0,\omega]$, *with* $(T^{2k+1}(R))$ *increasing and* $(T^{2k}(R))$ *decreasing.*

Proof. By Theorem 4.7, there exists at least one ω-periodic solution in $[a,R]$. Let $u \in [a,R]$ be any ω-periodic solution of (11.17). Since $f(t,z)$ is decreasing in z on $[a,R]$, from $a \leq u \leq R$, we obtain

$$a \leq T(R) \leq T(u) = u.$$

Then

$$a \leq T(R) \leq u \leq T^2(R).$$

By (11.27) this yields

$$a \leq T(R) \leq T^3(R) \leq u \leq T^2(R) \leq R.$$

We successively obtain

$$\begin{aligned} a &\leq T(R) \leq T^3(R) \\ &\vdots \\ &\leq T^{2k+1}(R) \\ &\vdots \\ &\leq u \\ &\vdots \\ &\leq T^{2k}(R) \\ &\vdots \\ &\leq T^4(R) \leq T^2(R) \leq R. \end{aligned} \tag{11.28}$$

Since T is completely continuous, there are two subsequences of $\left(T^{2k+1}(R)\right)$ and $\left(T^{2k}(R)\right)$ uniformly convergent to some $u^* \in [a, R]$ and $v^* \in [a, R]$, respectively. By (11.28) we see that the entire sequences $\left(T^{2k+1}(R)\right)$ and $\left(T^{2k}(R)\right)$ converge uniformly to u^*, respectively to v^*, and

$$a \leq u^* \leq u \leq v^* \leq R.$$

Obviously

$$u^* = T(v^*), \quad v^* = T(u^*).$$

Finally, use a similar argument as in the proof of Theorem 11.3 to conclude that $u^* = u = v^*$. ■

Theorem 11.14 *Assume the assumptions of Theorem 11.13 hold with*

$$T(a) \leq R \tag{11.29}$$

instead of (11.27). *Then* (11.17) *has a unique ω-periodic solution $u^* \in [a, R]$ and $T^k(w) \to u^*$ uniformly on $[0, \omega]$, for any $w \in [a, R]$.*

Proof. Notice (11.29) implies (11.27). Indeed, from $a \leq T(R) \leq T(a)$, we obtain

$$T^2(a) \leq T^2(R) \leq T(a) \leq R,$$

whence (11.27). Thus, Theorem 11.13 applies.

Further, for any $w \in [a, R]$ we have

$$a \leq T(R) \leq T(w) \leq T(a) \leq R.$$

This successively yields

$$\begin{aligned} a &\leq T(R) \leq T^3(R) \\ &\vdots \\ &\leq T^{2[(k-1)/2]+1}(R) \\ &\leq T^k(w) \\ &\leq T^{2[k/2]}(R) \\ &\vdots \\ &\leq T^4(R) \leq T^2(R) \leq R. \end{aligned}$$

Since $T^k(R) \to u^*$, it follows that $T^k(w) \to u^*$, as we wished. ∎

Remark 11.3 A sufficient condition for (11.29) is that

$$f(t,a) \leq \tau^{-1}R, \quad t \in \mathbf{R}.$$

For the next result let us replace (h4) by the following condition used in Guo–Lakshmikantham [28]:

(h4') $f(t,z) \leq \tau^{-1}R$ for all $t \in [0,\omega]$, $z \in [a,R]$.

The next theorems complement the results established in [28].

Theorem 11.15 *Assume (h1)–(h3) and (h4′) are satisfied. In addition assume $a > 0$, $f(t,z)$ is increasing in z on $[a,R]$, and there is a function $\varphi : [a/R,1) \to \mathbf{R}$ satisfying (11.23) for all $t \in [0,\omega]$, $\lambda \in [a/R,1)$ and $z \in \mathbf{R}^n$ with $z \in [a,R]$ and $\lambda z \geq a$. Then (11.17) has a unique ω-periodic solution $u^* \in [a,R]$ and $T^k(w) \to u^*$ uniformly on $[0,\omega]$, for any $w \in [a,R]$.*

Theorem 11.16 *Assume (h1)–(h3) and (h4′) are satisfied. In addition assume $a > 0$, $f(t,z)$ is decreasing in z on $[a,R]$, and there is a function $\chi : [a/R,1) \to (0,\infty)$ satisfying (11.24) for all $t \in [0,\omega]$, $\lambda \in [a/R,1)$ and $z \in \mathbf{R}^n$ with $z \in [a,R]$ and $\lambda z \geq a$. Then (11.17) has a unique ω-periodic solution $u^* \in [a,R]$ and $T^k(w) \to u^*$ uniformly on $[0,\omega]$, for any $w \in [a,R]$.*

The proofs are similar to that of Theorem 11.13, so we omit the details.

The monotone iterative method for periodic solutions to equation (11.17), for the case when $f(t,z)$ is increasing in z was discussed by Guo – Lakshmikantham [28]. See also Cañada [11], Cañada–Zertiti [12] and Guo – Lakshmikantham – Liu [29].

For $\chi(\lambda) = \lambda^{\alpha}$, $\alpha \in (-1,0)$, and $n = 1$, Theorems 11.13 and 11.15 were established in Precup [46], where several examples can also be found.

11.5 Methods of Upper and Lower Solutions for Equations of Hammerstein Type

In this section monotone iterative methods are used to localize and approximate solutions of the abstract Hammerstein equation in $\mathbf{R}^n$

$$u(x) = AN_f(u)(x) \quad \text{a.e. on } \Omega. \tag{11.30}$$

Here N_f is Nemytskii's superposition operator associated to a given function $f : \Omega \times \mathbf{R}^n \to \mathbf{R}^n$ ($\Omega \subset \mathbf{R}^N$ open), and A is a bounded linear operator from $L^q(\Omega;\mathbf{R}^n)$ to $L^p(\Omega;\mathbf{R})$. We seek solutions u in an order interval $[u_0, v_0]$ of $L^p(\Omega;\mathbf{R}^n)$, where u_0 is a lower solution of (11.30) and v_0 is a upper solution of (11.30). Of course, we shall assume that N_f maps $[u_0, v_0]$ into $L^q(\Omega;\mathbf{R}^n)$. The main assumption is the monotonicity of f in its second argument.

The advantage of working in the space $L^p(\Omega;\mathbf{R}^n)$ $(1 \leq p < \infty)$ ordered by the regular cone of positive functions is that we do not need the complete continuity of the operator $T = AN_f$.

Theorem 11.17 *Let $p, q \in [1,\infty)$, $A : L^q(\Omega;\mathbf{R}^n) \to L^p(\Omega;\mathbf{R}^n)$ a bounded linear operator and $f : \Omega \times \mathbf{R}^n \to \mathbf{R}^n$ a (p,q)-Carathéodory function. Let $u_0, v_0 \in L^p(\Omega;\mathbf{R}^n)$, $u_0 \leq v_0$, be such that*

$$u_0 \leq AN_f(u_0), \quad AN_f(v_0) \leq v_0. \tag{11.31}$$

Assume that

$$f(x,z) \leq f(x,z') \tag{11.32}$$

for all $z, z' \in \mathbf{R}^n$ with $u_0(x) \leq z \leq z' \leq v_0(x)$, a.e. on Ω, and that A is increasing, i.e.,

$$A(v) \geq 0 \quad \text{for all } v \geq 0. \tag{11.33}$$

Then there are solutions $u^*, v^* \in [u_0, v_0]$ to (11.30), minimal, respectively, maximal in $[u_0, v_0]$. Moreover, the sequences $(T^k(u_0))$, $(T^k(v_0))$ converge monotonically to u^* and v^*, respectively.

Proof. According to Theorem 5.1 N_f is well defined, bounded, and continuous from $L^p(\Omega;\mathbf{R}^n)$ to $L^q(\Omega;\mathbf{R}^n)$. From (11.32) it follows that N_f is increasing on $[u_0, v_0]$. This together with (11.33) guarantees that $T = AN_f$ is increasing on $[u_0, v_0]$. In addition T is continuous and

$$T([u_0, v_0]) \subset [u_0, v_0]$$

by (11.31). The conclusion now follows from Theorem 11.1 since the positive cone of $L^p(\Omega;\mathbf{R}^n)$ is regular. ■

Theorem 11.18 *Assume all the assumptions of Theorem 11.17 hold. In addition assume that there are constants* $a, b > 0$ *such that*

$$a \le u_0(x) \le v_0(x) \le b \tag{11.34}$$

and (11.32) *holds for all* $z, z' \in \mathbf{R}^n$ *with* $0 \le z \le z' \le v_0(x)$, *a.e. on* Ω. *Also assume that there is a function* $\phi : [a/b, 1) \to \mathbf{R}$ *with*

$$\phi(\lambda) > \lambda, \quad \phi(\lambda) f(x, z) \le f(x, \lambda z) \tag{11.35}$$

for all $\lambda \in [a/b, 1)$ *and* $z \in \mathbf{R}^n$ *with* $u_0(x) \le z \le v_0(x)$, *a.e. on* Ω. *Then* (11.30) *has a unique solution* u^* *in* $[u_0, v_0]$, *and*

$$T^k(w) \to u^* \text{ in } L^p(\Omega;\mathbf{R}^n) \text{ for all } w \in [u_0, v_0].$$

Moreover, the sequences $(T^k(u_0))$ *and* $(T^k(v_0))$ *converge monotonically to* u^*.

Proof. Let u^*, v^* be the minimal and maximal solutions given by Theorem 11.17. Clearly $u^* \le v^*$. We have to prove that $u^* = v^*$. Let

$$\lambda = \sup\{\mu \in [0,1] : \mu v_i^*(x) \le u_i^*(x) \text{ a.e. on } \Omega,\ i = 1, 2, ..., n\}. \tag{11.36}$$

From (11.34),

$$\frac{a}{b} \le \frac{u_i^*(x)}{v_i^*(x)} \le 1 \text{ a.e. on } \Omega,$$

so $\lambda \in [a/b, 1]$. Clearly, $\lambda v^* \le u^*$. If $\lambda < 1$, then using (11.35) we obtain

$$\phi(\lambda) v^* = \phi(\lambda) T(v^*) \le T(\lambda v^*) \le T(u^*) = u^*.$$

Since $\phi(\lambda) > \lambda$ this contradicts the definition of λ. Thus $\lambda = 1$ and so $u^* = v^*$. ■

A variant of Theorem 11.18 can be stated for the case where T is increasing on $[u_0, (b/a) u_0]$ instead of $[0, v_0]$.

Theorem 11.19 *Assume all the assumptions of Theorem 11.17 hold. In addition assume that there are constants* $a, b > 0$ *satisfying* (11.34) *such that* (11.32) *holds for all* $z, z' \in \mathbf{R}^n$ *with*

$$u_0(x) \leq z \leq z' \leq \frac{b}{a} u_0(x),$$

a.e. on Ω. *Also assume that there is a function* $\phi : [a/b, 1) \to \mathbf{R}$ *satisfying* (11.35) *for all* $\lambda \in [a/b, 1)$ *and* $z \in \mathbf{R}^n$ *with* $u_0(x) \leq z \leq (b/a) u_0(x)$ *and* $u_0(x) \leq \lambda z$, *a.e. on* Ω. *Then* (11.30) *has a unique solution* u^* *in* $[u_0, v_0]$, *and*

$$T^k(w) \to u^* \text{ in } L^p(\Omega; \mathbf{R}^n) \text{ for all } w \in [u_0, v_0].$$

Moreover, the sequences $(T^k(u_0))$, $(T^k(v_0))$ *converge monotonically to* u^*.

Proof. Let u^*, v^* be the minimal and maximal solutions given by Theorem 11.17. Let $\lambda \in [a/b, 1]$ be given by (11.36). Let $u = (u_1, u_2, ..., u_n)$ and $v = (v_1, v_2, ..., v_n)$ be defined by

$$u_i(x) = \max\{u_{0i}(x), \lambda v_i^*(x)\} = \lambda \max\left\{\frac{1}{\lambda} u_{0i}(x), v_i^*(x)\right\}$$

and

$$v_i(x) = \max\left\{\frac{1}{\lambda} u_{0i}(x), v_i^*(x)\right\},$$

$i = 1, 2, ..., n$. We have

$$v^* \leq v \leq \frac{b}{a} u_0, \quad u_0 \leq u = \lambda v \leq u^*.$$

Hence if $\lambda < 1$ then

$$\phi(\lambda) v^* = \phi(\lambda) T(v^*) \leq \phi(\lambda) T(v) \leq T(\lambda v) \leq T(u^*) = u^*.$$

Since $\phi(\lambda) > \lambda$ this contradicts the definition of λ. Thus $\lambda = 1$ and so $u^* = v^*$. ∎

Similar existence and uniqueness results can be established for decreasing operators. For example, we have the following theorem.

Theorem 11.20 *Let* $p, q \in [1, \infty)$, $A : L^q(\Omega; \mathbf{R}^n) \to L^p(\Omega; \mathbf{R}^n)$ *an increasing bounded linear operator and* $f : \Omega \times \mathbf{R}^n \to \mathbf{R}^n$ *a* (p, q)*-Carathéodory function. Let* $u_0, v_0 \in L^p(\Omega; \mathbf{R}^n)$ *and* $a, b > 0$ *be such that* (11.34) *and*

$$u_0 \leq AN_f(v_0), \quad AN_f(u_0) \leq v_0 \tag{11.37}$$

hold. Assume that

$$f\left(x,z'\right) \leq f\left(x,z\right) \tag{11.38}$$

for all $z, z' \in \mathbf{R}^n$ *with* $0 \leq z \leq z' \leq v_0(x)$, *a.e. on* Ω. *Also assume that there is a function* $\chi : [a/b, 1) \to (0, \infty)$ *with*

$$\chi(\lambda) < \frac{1}{\lambda}, \quad f(t, \lambda z) \leq \chi(\lambda) f(t, z) \tag{11.39}$$

for all $\lambda \in [a/b, 1)$ *and* $z \in \mathbf{R}^n$ *with* $u_0(x) \leq z \leq v_0(x)$, *a.e. on* Ω. *Then there exists a unique solution* $u^* \in [u_0, v_0]$ *to (11.30) and*

$$T^k(w) \to u^* \text{ in } L^p(\Omega; \mathbf{R}^n) \text{ for all } w \in [u_0, v_0].$$

Moreover, the sequences

$$\left(T^{2k}(u_0)\right), \left(T^{2k+1}(u_0)\right), \left(T^{2k}(v_0)\right), \left(T^{2k+1}(v_0)\right)$$

converge monotonically to u^*.

Proof. The operator T is continuous and decreasing on $[u_0, v_0]$. In addition $T([u_0, v_0]) \subset [u_0, v_0]$, $\left(T^{2k}(u_0)\right)$ is increasing, $\left(T^{2k+1}(u_0)\right)$ is decreasing, and $T^{2k}(u_0) \leq T^{2k+1}(u_0)$. The positive cone of $L^p(\Omega; \mathbf{R}^n)$ being regular, the two sequences converge to some elements u^* and v^*, respectively. Then

$$u_0 \leq u^* \leq v^* \leq v_0, \quad T(u^*) = v^*, \quad T(v^*) = u^*.$$

To prove $u^* = v^*$ let $\lambda \in [a/b, 1]$ be given by (11.36). Then if $\lambda < 1$, from

$$0 \leq \lambda v^* \leq u^* \leq v_0$$

using the monotonicity of T on $[0, v_0]$ and (11.39) we deduce

$$v^* = T(u^*) \leq T(\lambda v^*) \leq \chi(\lambda) T(v^*) = \chi(\lambda) u^*,$$

where $\chi(\lambda) < 1/\lambda$, a contradiction to the definition of λ. Thus $\lambda = 1$, and so $u^* = v^*$. ■

Our next goal is to obtain lower and upper solutions. We succeed this if we have more information about f and A, in particular, about the spectrum of A.

Theorem 11.21 *Let $p, q \in [1, \infty)$, $A : L^q(\Omega; \mathbf{R}^n) \to L^p(\Omega; \mathbf{R}^n)$ an increasing linear operator and $f : \Omega \times \mathbf{R}^n \to \mathbf{R}^n$ a (p,q)-Carathéodory function. Assume that there are $c \in \mathbf{R}_+$ and $g \in L^q(\Omega; \mathbf{R}^n_+)$ such that*

$$f(x, z) \leq cz + g(x) \tag{11.40}$$

for a.e. $x \in \Omega$ and all $z \geq 0$. Then any solution $v_0 \geq 0$ of the equation

$$(I - cA)(v) = A(g) \tag{11.41}$$

(if there is one) is an upper solution of the equation $u = AN_f(u)$. If in addition

$$-f(x, -z) \leq cz + g(x) \tag{11.42}$$

for a.e. $x \in \Omega$ and all $z \geq 0$, then $u_0 := -v_0$ is a lower solution.

Proof. Assume $v_0 \geq 0$ solves (11.41). Then, from (11.40) we have

$$f(x, v_0(x)) \leq c\, v_0(x) + g(x)$$

and since A is increasing,

$$AN_f(v_0) \leq cA(v_0) + A(g) = v_0.$$

Hence v_0 is a upper solution. We leave to the reader to check that $-v_0$ is a lower solution. ∎

We conclude this section by two much applicable results involving spectral properties of A.

First we establish an *abstract Poincaré inequality.*

Lemma 11.1 *Let X be a Hilbert space and $A : X \to X$ be a positive self-adjoint operator. Then*

$$|A(u)|^2 \leq |A|(A(u), u), \quad u \in X. \tag{11.43}$$

Proof. Since A is positive, for all $u, v \in X$ and $t \in \mathbf{R}$, we have

$$(A(u + tv), u + tv) \geq 0,$$

that is

$$(A(v), v)t^2 + 2(A(u), v)t + (A(u), u) \geq 0.$$

Consequently

$$(A(u), v)^2 \leq (A(v), v)(A(u), u).$$

For $v = A(u)$ this inequality becomes

$$|A(u)|^4 \leq (A^2(u), A(u))(A(u), u). \tag{11.44}$$

According to Proposition 6.2,

$$(A^2(u), A(u)) \leq |A|\,|A(u)|^2. \tag{11.45}$$

Now (11.44) and (11.45) yield (11.43). ■

Our next result is an *abstract weak maximum principle* in $L^2(\Omega; \mathbf{R}^n)$. For a function $u : \Omega \to \mathbf{R}^n$, we let u^+, u^- be the functions defined by

$$u_i^+(x) = \max\{0, u_i(x)\}, \quad u_i^-(x) = \max\{0, -u_i(x)\},$$

$i = 1, 2, ..., n$. Clearly, $u = u^+ - u^-$, $u^+ \geq 0$ and $u^- \geq 0$. Also, for a function u one has $u \geq 0$, if and only if $u^- = 0$.

Lemma 11.2 *Let* $A : L^2(\Omega; \mathbf{R}^n) \to L^2(\Omega; \mathbf{R}^n)$ *be a positive self-adjoint operator. Assume the following conditions are satisfied:*

$$A(u) \geq 0 \quad \text{for } u \geq 0; \quad A(u) \neq 0 \quad \text{for } u \neq 0 \tag{11.46}$$

and

$$(A(u^+), A(u^-))_2 = (A(u^+), u^-)_2 = 0, \quad u \in L^2(\Omega; \mathbf{R}^n). \tag{11.47}$$

Then for any constant $c < |A|^{-1}$,

$$(I - cA)^{-1}(u) \geq 0 \quad \text{for all } u \geq 0. \tag{11.48}$$

Proof. Let $\sigma(A)$ be the spectrum of A, that is

$$\sigma(A) = \mathbf{R} \setminus \{\lambda \in \mathbf{R} : A - \lambda I \text{ is bijective}\}.$$

It is known that

$$\sigma(A) \subset [-|A|, |A|]$$

(see Brezis [7], p 94). Since $c < |A|^{-1}$ the operator $I - cA$ is invertible. Let $u \geq 0$ and let $v = (I - cA)^{-1}(u)$. Clearly

$$v - cA(v) = u. \tag{11.49}$$

We have to show that $v \geq 0$, equivalently $v^- = 0$. Assume the contrary, i.e., $v^- \neq 0$. Then (11.46) guarantees that $A(v^-) \geq 0$ and $A(v^-) \neq 0$. If we multiply (11.49) by $A(v^-)$, and we use (11.47), we obtain

$$-\left(A\left(v^-\right), v^-\right)_2 + c\left(A\left(v^-\right), A\left(v^-\right)\right)_2 = \left(A\left(v^-\right), u\right)_2.$$

Since both u and $A(v^-)$ are positive we have

$$\left(A\left(v^-\right), u\right)_2 \geq 0.$$

Therefore

$$c \geq \frac{\left(A\left(v^-\right), v^-\right)_2}{\left|A\left(v^-\right)\right|_2^2}.$$

This together with (11.43) implies $c \geq |A|^{-1}$, a contradiction. Thus $v^- = 0$ and the proof is complete. ■

For the last two results assume that $\Omega \subset \mathbf{R}^N$ is bounded open.

Theorem 11.22 *Let $A : L^2(\Omega; \mathbf{R}^n) \to L^2(\Omega; \mathbf{R}^n)$ be a positive self-adjoint operator such that* (11.46) *and* (11.47) *hold. Let $f : \Omega \times \mathbf{R}^n \to \mathbf{R}^n$ be a map satisfying the Carathéodory conditions, such that for each $m \in (0, \infty)$ there is a constant $a_m \in \mathbf{R}_+$ with*

$$f(x, z) + a_m z \quad \text{increasing in } z \text{ on } [-m, m]$$

for a.e. $x \in \Omega$. Assume that

$$f(x, z) \leq cz + c', \quad f(x, -z) \geq -cz - c' \tag{11.50}$$

for a.e. $x \in \Omega$, all $z \in \mathbf{R}^n$ with $z \geq 0$, and some $c \in \mathbf{R}_+$ with $c < |A|^{-1}$ and $c' \in \mathbf{R}^n_+$. In addition assume that the solution of the equation

$$u - cA(u) = A(c')$$

belongs to $L^\infty(\Omega; \mathbf{R}^n)$. Then the equation $u = AN_f(u)$ has at least one solution in $L^2(\Omega; \mathbf{R}^n)$. Moreover, if the set S_+ (S_-) of all solutions $u \geq 0$ (respectively, $u \leq 0$) is nonempty, then it has a maximal (respectively, minimal) element.

Proof. Since $c < |A|^{-1}$, the operator $I - cA$ is bijective and so the equation (11.41) has a unique solution v_0 for each g. Here $g = c'$. By Lemma 11.2, $v_0 \geq 0$. Now, (11.50) guarantees both (11.40), (11.42). Thus, by Theorem 11.21, v_0 is an upper solution and $u_0 = -v_0$ is a lower solution.

Since v_0 belongs to $L^\infty(\Omega; \mathbf{R}^n)$, there is $m \in (0, \infty)$ with $v_0 \le m$. Then the function f_m given by

$$f_m(x, z) = f(x, z) + a_m z$$

is increasing in z on $[-m, m]$. Also the equation $u = AN_f(u)$ is equivalent to

$$u = (I + a_m A)^{-1} AN_{f_m}(u).$$

Let

$$T_m = (I + a_m A)^{-1} AN_{f_m}.$$

Clearly,

$$u_0 \le T_m(u_0), \quad T_m(v_0) \le v_0$$

and T_m is continuous and increasing on $[u_0, v_0]$. Let u^*, v^* be the minimal, respectively maximal solution in $[u_0, v_0]$. We have

$$-v_0 \le u^* \le v^* \le v_0.$$

We now show that if $w \in L^2(\Omega; \mathbf{R}^n)$, $w \ge 0$, solves $w = AN_f(w)$ then $w \le v_0$. Indeed, from

$$w = AN_f(w) \le A(cw + c') = cA(w) + A(c')$$

and

$$v_0 = cA(v_0) + A(c'), \tag{11.51}$$

by subtraction we obtain

$$v_0 - w \ge cA(v_0 - w).$$

Then by the maximum principle, Lemma 11.2, $v_0 - w \ge 0$. Hence v^* is maximal in $\mathcal{S}_+$. Similarly, if $w \in L^2(\Omega)$, $w \le 0$ and $w = AN_f(w)$, then $-v_0 \le w$. Hence u^* is minimal in $\mathcal{S}_-$. ■

The last theorem is an existence and localization result of a nonnegative non-zero solution.

Theorem 11.23 *Let* $A : L^2(\Omega; \mathbf{R}^n) \to L^2(\Omega; \mathbf{R}^n)$ *be a completely continuous positive self-adjoint operator such that* (11.46) *and* (11.47) *hold. Let* $f : \mathbf{R}^n_+ \to \mathbf{R}^n$ *be a continuous map such that* $f(0) = 0$ *and for each* $m \in (0, \infty)$, *there is a constant* $a_m \in \mathbf{R}_+$ *with*

$$f_m(z) := f(z) + a_m z \ \textit{increasing on} \ [0, m].$$

Assume that

$$f(z) \le cz + c'$$

for all $z \in \mathbf{R}_+^n$*, and some* $c \in \mathbf{R}_+$ *with* $c < |A|^{-1}$, $c' \in (0, \infty)^n$, *and*

$$f(z) \ge |A|^{-1} z \tag{11.52}$$

for all $z \in \mathbf{R}_+^n$ *with* $|z| \le \varepsilon_0$*, where* $\varepsilon_0 > 0$*. In addition assume that the solutions of the equations*

$$u - cA(u) = A(c') \quad \text{and} \quad u - |A|^{-1} A(u) = 0$$

belong to $L^\infty(\Omega; \mathbf{R}^n)$*. Then the equation* $u = AN_f(u)$ *has a maximal solution* u *in* $L^2(\Omega; \mathbf{R}_+^n)$ *and* $u \neq 0$.

Proof. As above, the unique solution v_0 of the equation

$$u - cA(u) = A(c')$$

belongs to $L^\infty(\Omega; \mathbf{R}_+^n)$ and is an upper solution of the equation $u = AN_f(u)$. Since $f(0) = 0$ the null function is a solution, and so a lower solution. Now we apply Theorem 11.1 to deduce the existence of a maximal fixed point v^* in $[0, v_0]$ of the operator

$$T_m = (I + a_m A)^{-1} AN_{f_m},$$

where $m \in (0, \infty)$ satisfies $v_0(x) \le m$ a. e. on Ω. As in the proof of Theorem 11.22 we can show that v^* is maximal in the set of all nonnegative solutions. To show that $v^* \neq 0$, we prove that v^* is the maximal fixed point of T_m in an order subinterval

$$[u_0, v_0] \subset [0, v_0]$$

with $u_0 \neq 0$.

Since A is completely continuous, the supremum in (6.3) is attained at some u_1 with $|u_1|_2 = 1$. Furthermore, since A is positive, we may assume that

$$|A| = (A(u_1), u_1)_2 .$$

Then, according to (11.47), we have

$$\begin{aligned} |A| &= \left(A\left(u_1^+ - u_1^-\right), u_1^+ - u_1^-\right)_2 \\ &= \left(A\left(u_1^+\right), u_1^+\right)_2 + \left(A\left(u_1^-\right), u_1^-\right)_2 \\ &= \left(A\left(u_1^+ + u_1^-\right), u_1^+ + u_1^-\right)_2 . \end{aligned}$$

Hence we may assume that $u_1 \geq 0$ (otherwise, replace u_1 by $u_1^+ + u_1^-$; notice that $\left|u_1^+ + u_1^-\right|_2 = \left|u_1^+ - u_1^-\right|_2 = |u_1|_2 = 1$). For any fixed $v \in L^2(\Omega; \mathbf{R}^n)$ we consider the function

$$g(t) = \frac{(A(u_1 + tv), u_1 + tv)_2}{|u_1 + tv|_2^2},$$

which can be defined on a neighborhood of $t = 0$. This function attains its maximum $|A|$ at $t = 0$, so $g'(0) = 0$. Notice

$$g'(0) = 2\left[(A(u_1), v)_2 - |A|(u_1, v)_2\right].$$

Hence

$$u_1 = |A|^{-1} A(u_1)$$

(i.e., $|A|$ is the largest eigenvalue of A and u_1 is an eigenfunction). Also, by hypothesis u_1 belongs to $L^\infty(\Omega; \mathbf{R}^n)$. Let

$$u_0 = \varepsilon |u_1|_\infty^{-1} u_1,$$

where $0 < \varepsilon \leq \varepsilon_0$. Clearly

$$u_0 \geq 0, \quad u_0 \neq 0, \quad |u_0(x)| \leq \varepsilon \text{ a.e. on } \Omega, \quad u_0 = |A|^{-1} A(u_0).$$

Using (11.52), we deduce

$$\begin{aligned} u_0 &= |A|^{-1} A(u_0) = A\left(|A|^{-1} u_0\right) \\ &\leq AN_f(u_0). \end{aligned}$$

Thus u_0 is a lower solution of $u = AN_f(u)$. Also, from

$$v_0 = cA(v_0) + A(c'), \quad u_0 = |A|^{-1} A(u_0),$$

we have

$$v_0 - u_0 = cA(v_0 - u_0) + \left(c - |A|^{-1}\right) A(u_0) + A(c').$$

Now we choose $\varepsilon > 0$ small enough so that

$$\left(c - |A|^{-1}\right) u_0(x) + c' \geq 0 \quad \text{a.e. on } \Omega.$$

Then

$$v_0 - u_0 - cA(v_0 - u_0) \geq 0,$$

and by the maximum principle, $v_0 - u_0 \geq 0$. Next we apply Theorem 11.1 to deduce the existence of a maximal fixed point in $[u_0, v_0]$ of T_m. Clearly it is equal to v^*. ■

Example 11.5 Let $n = 1$. The operator

$$A = (-\Delta)^{-1}$$

has all the properties required by Theorems 11.22–11.23. Moreover, in this case u^*, v^* are, respectively, the minimal and maximal solutions in the set of all solutions in $L^2(\Omega)$. Indeed, if $w \in L^2(\Omega)$ is any solution and we let f_w be defined by

$$f_w(x,z) = \begin{cases} f(x,z) & \text{if } w(x) > 0, \\ 0 & \text{if } w(x) \le 0, \end{cases}$$

then

$$-\Delta w^+ = f_w(x, w^+) \le c\,w^+(x) + c'$$

a.e. on Ω. Hence

$$w^+ \le cA(w^+) + A(c').$$

This together with (11.51) implies

$$v_0 - w^+ \ge cA(v_0 - w^+).$$

Thus $w^+ \le v_0$. Similarly $-v_0 \le -w^-$. Therefore $-v_0 \le w \le v_0$.

For other applications of the monotone iterative technique to different types of equations we refer the reader to Bainov–Hristova [4], Buică [8], Carl–Heikkilä [13], Constantin [15], De Coster–Habets [19], Fabry–Habets [22], Liz–Nieto [35], Pouso [45], Precup [49] and Wang–Cabada–Nieto [61].

Chapter 12

Quadratically Convergent Methods

The succesive approximation method as well as the monotone iterative methods described previously are not very fast. To explain this let us consider an operator $T : X \to X$. All these methods give us convergent sequences (u_k) of the form

$$u_{k+1} = T(u_k), \quad k \in \mathbf{N}$$

having as limit a fixed point u^* of T. For the results in Chapter 10, we have

$$d(u_{k+1}, u^*) \leq M d(u_k, u^*).$$

This shows that the convergence is linear.

In Chapter 11 we have not assumed that T was a contraction. However, if T is Lipschitz in a neighborhood of u^*, with the Lipschitz constant c, then we have

$$\begin{aligned} |u_{k+1} - u^*| &= |T(u_k) - T(u^*)| \\ &\leq c\,|u_k - u^*| \end{aligned}$$

for all large k, which shows that the convergence is again linear.

Faster iterative methods for solving nonlinear equations can be derived from the well known Newton's method. In this short chapter we describe just Newton's method and we explain the way it can be used to solve nonlinear integral equations. When Newton's method is applied to the differential equations theirselves, one speaks about the *quasilinearization* method. This method offers iterative sequences of approximate solutions that converge quadratically to the solution. We say that the convergence of the sequence (u_k) to u^* is of *order* p if

$$|u_{k+1} - u^*| \leq c\,|u_k - u^*|^p$$

for all large k and some $c > 0$. When $p = 2$ we say that the convergence is *quadratic.* In case that the nonlinear terms of the operator equations have behavior properties of convexity type, the iterative sequences of approximate solutions are also monotone.

12.1 Newton's Method

In this section Newton's method is presented following Deimling [18], Section 15.4 and Granas–Guenther–Lee [26]. For more information, we send to Jankó [31], Kantorovitch–Akilov [32], Moore [36] and Potra–Rheinboldt [44].

Let X, Y be Banach spaces, $U \subset X$ an open set and let $F : U \to Y$ be continuously Fréchet differentiable. *Newton's method* is an iterative method for solving

$$F(u) = 0.$$

It works as follows: choose an initial approximation $u_0 \in U$ to the solution and generate a sequence (u_k) by solving successively, if possible, the equations

$$F(u_k) + F'(u_k)(u_{k+1} - u_k) = 0, \quad k \in \mathbf{N} \tag{12.1}$$

If (u_k) converges to a point $u^* \in U$, then by (12.1), $F(u^*) = 0$.

Notice that if $F'(u)^{-1}$ exists for all $u \in U$, then (u_k) is the sequence of successive approximations for the equation

$$u = G(u),$$

where

$$G(u) = u - F'(u)^{-1} F(u).$$

Hence

$$u_{k+1} = u_k - F'(u_k)^{-1} F(u_k), \quad k \in \mathbf{N}. \tag{12.2}$$

The following theorem contains the convergence criterion for Newton's method.

Theorem 12.1 (Kantorovitch) *Let X, Y be Banach spaces and $F : B_r(u_0) \subset X \to Y$ a C^1-operator. Assume:*

(*i*) $F'(u_0)^{-1} \in L(Y, X)$, $\left|F'(u_0)^{-1} F(u_0)\right| \leq \alpha$ *and* $\left|F'(u_0)^{-1}\right| \leq \beta$;

(*ii*) $|F'(u) - F'(v)| \leq \gamma |u - v|$ *for all* $u, v \in B_r(u_0)$;

(*iii*) $2\alpha\beta\gamma < 1$ *and* $2\alpha < r$.

Then F *has a unique zero* u^* *in* $\overline{B}_{2\alpha}(u_0)$ *and the sequence* (u_k) *given by* (12.1) *converges quadratically to* u^* *and satisfies*

$$|u_k - u^*| \leq \alpha 2^{1-k} (2\alpha\beta\gamma)^{2^k - 1}, \quad k \in \mathbf{N}. \tag{12.3}$$

Proof. First we prove that $F'(u)$ is invertible for any $u \in \overline{B}_{2\alpha}(u_0)$. For this, let $u \in \overline{B}_{2\alpha}(u_0)$ and define the map $\Gamma : X \to X$ as

$$\Gamma = I - F'(u_0)^{-1} F'(u) = F'(u_0)^{-1}(F'(u_0) - F'(u)).$$

Then for every $v, w \in X$ we have

$$\begin{aligned} |\Gamma(v) - \Gamma(w)| &\leq \left|F'(u_0)^{-1}\right| |F'(u_0) - F'(u)| \, |v - w| \\ &\leq \beta\gamma |u_0 - u| \, |v - w| \leq 2\alpha\beta\gamma |v - w|. \end{aligned}$$

Since $2\alpha\beta\gamma < 1$, Γ is a contraction. Consequently the map

$$I - \Gamma = F'(u_0)^{-1} F'(u)$$

is bijective. As a result

$$F'(u) = F'(u_0)(I - \Gamma)$$

is bijective too.

Next we prove that the sequence (u_k) given by (12.2) is well defined and convergent.

We shall prove by induction that for each $k \in \mathbf{N} \setminus \{0\}$,

$$|u_k - u_0| \leq 2\alpha, \quad \alpha_k \leq 2^{-k+1}\alpha, \quad \gamma_k \leq \frac{1}{2}, \tag{12.4}$$

where

$$\gamma_k = \gamma\alpha_k\beta_k, \quad \alpha_k = \left|F'(u_{k-1})^{-1} F(u_{k-1})\right|, \quad \beta_k = \left|F'(u_{k-1})^{-1}\right|.$$

Assume (12.4) holds for 1, 2, ..., k. We shall prove that (12.4) is also true for $k+1$. We have

$$\begin{aligned} & F'(u_k)^{-1} F(u_k) \\ = \; & F'(u_k)^{-1} \left[F(u_k) - F(u_{k-1}) - F'(u_{k-1})(u_k - u_{k-1})\right] \\ = \; & F'(u_k)^{-1} \int_0^1 \left[F'(u_{k-1} + t(u_k - u_{k-1})) - F'(u_{k-1})\right](u_k - u_{k-1}) \, dt. \end{aligned}$$

Then from (ii) we obtain

$$\begin{aligned}\alpha_{k+1} &\leq \beta_{k+1}\alpha_k \int_0^1 \left|F'\left(u_{k-1}+t\left(u_k-u_{k-1}\right)\right)-F'\left(u_{k-1}\right)\right| dt \quad (12.5)\\ &\leq \gamma\beta_{k+1}\alpha_k^2 \int_0^1 t\,dt = \frac{1}{2}\gamma\beta_{k+1}\alpha_k^2.\end{aligned}$$

On the other hand, from

$$F'\left(u_k\right)=F'\left(u_{k-1}\right)\left[I+F'\left(u_{k-1}\right)^{-1}\left(F'\left(u_k\right)-F'\left(u_{k-1}\right)\right)\right],$$

we deduce that

$$\beta_{k+1} \leq \frac{\beta_k}{1-\gamma_k}.$$

Hence

$$\alpha_{k+1} \leq \frac{1}{2}\frac{\gamma\beta_k\alpha_k^2}{1-\gamma_k}.$$

Thus

$$\alpha_{k+1} \leq \frac{1}{2}\frac{\alpha_k\gamma_k}{1-\gamma_k}, \quad \gamma_{k+1} \leq \frac{1}{2}\frac{\gamma_k^2}{\left(1-\gamma_k\right)^2}. \tag{12.6}$$

Since $\gamma_k \leq 1/2$ we obtain

$$\gamma_{k+1} \leq \frac{1}{2}.$$

Then

$$\alpha_{k+1} \leq 2^{-1}\alpha_k \leq 2^{-k}\alpha.$$

In addition

$$\begin{aligned}\left|u_{k+1}-u_0\right| &\leq \left|u_{k+1}-u_k\right|+\left|u_k-u_{k-1}\right|+\ldots+\left|u_1-u_0\right|\\ &= \alpha_{k+1}+\alpha_k+\ldots+\alpha_1 \leq \left(2^{-k}+2^{-k+1}+\ldots+1\right)\alpha\\ &\leq 2\alpha.\end{aligned}$$

Hence (12.4) holds for all $k \in \mathbf{N}\setminus\{0\}$. Thus (u_k) is well defined and is a Cauchy sequence. Let $u^* \in \overline{B}_{2\alpha}(u_0)$ be its limit. Clearly $F(u^*)=0$. Now an inequality like (12.5) guarantees

$$\left|u_{k+1}-u^*\right| \leq c\left|u_k-u^*\right|^2$$

with $c=2^{-1}\gamma\sup_k \beta_k < \infty$ since $\beta_k \to \left|F'(u^*)^{-1}\right|$. Hence (u_k) converges quadratically to a zero of F. If $v^* \in \overline{B}_{2\alpha}(u_0)$ is another zero of F then from

$$\begin{aligned} v^* - u^* &= F'(u_0)^{-1}\left[F(u^*) - F(v^*) - F'(u_0)(u^* - v^*)\right] \\ &= F'(u_0)^{-1}\int_0^1 \left[F'(v^* + t(u^* - v^*)) - F'(u_0)\right](u^* - v^*)\,dt, \end{aligned}$$

we obtain

$$\begin{aligned} |u^* - v^*| &\leq \beta\gamma\,|u^* - v^*|\int_0^1 |v^* + t(u^* - v^*) - u_0|\,dt \\ &\leq 2\alpha\beta\gamma\,|u^* - v^*|. \end{aligned}$$

Thus $v^* = u^*$. Finally we prove (12.3). Let $\delta_k = \gamma_k(1 - \gamma_k)^{-1}$. Since $\gamma_k \leq 1/2$, (12.6) implies $\delta_k \leq \delta_{k-1}^2$. Hence $\delta_k \leq \delta_1^{2^{k-1}}$. Consequently

$$\alpha_k \leq 2^{-1}\delta_1^{2^{k-2}}\alpha_{k-1} \leq \ldots \leq 2^{-k+1}(2\alpha\beta\gamma)^{2^{k-1}-1}\alpha.$$

Therefore

$$|u_k - u_0| \leq \sum_{i=k+1}^{\infty}\alpha_i \leq \alpha 2^{1-k}(2\alpha\beta\gamma)^{2^k-1}.$$

The proof is complete. ■

Example 12.1 Let us present Newton's method for the abstract Hammerstein equation in R^n

$$u = AN_f(u), \quad u \in L^p(\Omega;\mathbf{R}^n).$$

We assume that $p, q \in [1,\infty)$, $p \geq q$ and the following conditions are satisfied:

(i) $f : \Omega \times R^n \to R^n$ is differentiable with respect to z and $\partial f_i/\partial z_j$ is $(p, pq/(p-q))$-Carathéodory for all $i, j \in \{1, 2, ..., n\}$.

(ii) $A : L^q(\Omega;\mathbf{R}^n) \to L^p(\Omega;\mathbf{R}^n)$ is bounded linear.

Let $F : L^p(\Omega;\mathbf{R}^n) \to L^p(\Omega;\mathbf{R}^n)$ be defined as

$$F(u) = u - AN_f(u).$$

Then F is a C^1-operator and

$$F'(u)(h) = h - A\left[N_{J_f}(u)\,h\right],$$

for all $u, h \in L^p(\Omega; \mathbf{R}^n)$, where

$$J_f = \left[\frac{\partial f_i}{\partial z_j}\right]_{1\le i,j\le n}$$

is the *Jacobian matrix* of f.

Now let $u_0 \in L^p(\Omega; \mathbf{R}^n)$ be an initial approximation of the solution. The *Newton iterates* are determined recursively by solving

$$F(u_k) + F'(u_k)(u_{k+1} - u_k) = 0,$$

that is, the linear equation

$$u_{k+1} - A\left[N_{J_f}(u_k)\, u_{k+1}\right] = A\left[N_f(u_k) - N_{J_f}(u_k)\, u_k\right].$$

Notice $F'(u)^{-1}$ exists if the only solution to the linear equation

$$h - A\left[N_{J_f}(u)\, h\right] = 0$$

is $h = 0$.

Example 12.2 Here we apply Newton's method to a differential equation itself. In this situation one speaks about the *quasilinearization method.* For the original idea of this method, see Bellman–Kalaba [6]. Consider the problem

$$\begin{cases} u'' = f(x, u, u'), & x \in [0,1] \\ u(0) = u(1) = 0. \end{cases}$$

Assume $f : [0,1] \times R^{2n} \to R^n$ is continuous and continuously differentiable with respect to $z \in R^{2n}$. Let $C_B^2([0,1]; \mathbf{R}^n)$ be the Banach space of all functions from $C^2([0,1]; \mathbf{R}^n)$ which satisfy the boundary conditions $u(0) = u(1) = 0$. Consider the operator $F : C_B^2([0,1]; \mathbf{R}^n) \to C([0,1]; \mathbf{R}^n)$, defined by

$$F(u) = u'' - f(., u, u').$$

We have that F is a C^1-operator with

$$F'(u)(h) = h'' - J_f^1(., u, u')\, h - J_f^2(., u, u')\, h'$$

for $h \in C_B^2([0,1]; \mathbf{R}^n)$. Here

$$\begin{aligned} J_f^1(x, z, z') &= \left[\frac{\partial f_i}{\partial z_j}(x, z, z')\right]_{1\le i,j\le n}, \\ J_f^2(x, z, z') &= \left[\frac{\partial f_i}{\partial z'_j}(x, z, z')\right]_{1\le i,j\le n}. \end{aligned}$$

The Newton iterates are here determined by solving in $C_B^2([0,1];\mathbf{R}^n)$ the linear equation

$$\begin{aligned} & u_{k+1}'' - J_f^2(.,u_k,u_k')\,u_{k+1}' - J_f^1(.,u_k,u_k')\,u_{k+1} \\ = \;& f(.,u_k,u_k') - J_f^2(.,u_k,u_k')\,u_k' - J_f^1(.,u_k,u_k')\,u_k. \end{aligned}$$

Here again $F'(u)^{-1}$ exists if the only solution in $C_B^2([0,1];\mathbf{R}^n)$ to the linear equation

$$h'' - J_f^2(.,u,u')\,h' - J_f^1(.,u,u')\,h = 0$$

is $h = 0$. For this, there are known sufficient conditions given in terms of f, J_f^1 and J_f^2 (see Granas–Guenther–Lee [26]).

Notice that in applications it is often difficult to find an initial point u_0 and its neighborhood $B_r(u_0)$ such that all the assumptions of Theorem 12.1 are satisfied. One could overcome this difficulty by asking a convexity property on the nonlinear operator in order to obtain the monotonicity and the convergence of Newton's iteratives with respect to a certain order structure. We refer the reader to Vandergraft [60] and Schmidt–Schneider [57] for Newton's method in ordered spaces, and to Bellman–Kalaba [6] and Mureşan–Trif [37] for applications to differential equations. The 1990s have brought up to date the subject by combining Newton's method with the method of upper and lower solutions together with the monotone iterative technique, and by the new idea of considering more general nonlinearities which can be represented as a difference of two convex functions. The resulting mixture is now known as the *generalized quasilinearization* method. It provides a tool to construct upper and lower sequences that converge monotonically and quadratically to a solution of the equation. For several applications of this method see Lakshmikantham–Leela–Sivasundaram [34], Carl–Lakshmikantham [14], Cabada–Nieto [10] and the references therein.

12.2 Generalized Quasilinearization for an Integral Equation with Delay

The results presented in this section was adapted from Precup [48]. We use the method of quasilinearization generalized in Lakshmikantham–Leela–Sivasundaram [34], for the quadratic, monotonic, bilateral approximation of a solution to the delay integral equation

$$u(t) = \int_{t-\tau}^{t} f(s,u(s))\,ds. \tag{12.7}$$

We deal with the initial value problem for (12.7) and look for a continuous solution $u(t)$ for $-\tau \leq t \leq t_1$, when

$$u(t) = \varphi(t), \quad -\tau \leq t \leq 0. \tag{12.8}$$

We assume

$$\varphi(0) = \int_{-\tau}^{0} f(s, \varphi(s))\, ds. \tag{12.9}$$

Under assumption (12.9), problem (12.7)–(12.8) is equivalent with the initial value problem

$$\begin{cases} u'(t) = f(t, u(t)) - f(t-\tau, \widetilde{u}(t-\tau)), & 0 < t \leq t_1, \\ u(0) = \varphi(0), \end{cases} \tag{12.10}$$

where $u \in C^1[0, t_1]$ and $\widetilde{u}(t) = \begin{cases} \varphi(t) & \text{for } -\tau \leq t \leq 0 \\ u(t) & \text{for } 0 < t \leq t_1 \end{cases}$.

We shall discuss a more general problem of type (12.10), namely

$$\begin{cases} u'(t) = f(t, u(t)) + g(t-\tau, \widetilde{u}(t-\tau)), & 0 < t \leq t_1, \\ u(0) = \varphi(0). \end{cases} \tag{12.11}$$

Before we state the main result of this section we present two useful lemmas. The first one can be proved using Corollary 3.1.2 in Piccinini–Stampacchia–Vidossich [41].

Lemma 12.1 *Let $u, v \in C^1[a,b]$ be such that $u \leq v$, and let $H(t,z)$ be continuous for $u(t) \leq z \leq v(t)$ and every $t \in [a,b]$. Assume that*

$$u'(t) \leq H(t, u(t)), \quad H(t, v(t)) \leq v'(t)$$

on $[a,b]$. Then for each $\alpha_0 \in [u(a), v(a)]$ there exists a solution $\alpha \in C^1[a,b]$ of the problem

$$\begin{cases} \alpha'(t) = H(t, \alpha(t)), & t \in [a,b], \\ \alpha(a) = \alpha_0 \end{cases}$$

such that $u \leq \alpha \leq v$.

Proof. For $u(t) < v(t)$ on $[a,b]$ and $u(a) < \alpha_0 < v(a)$ this result is Corollary 3.1.2 in Piccinini–Stampacchia–Vidossich [41]. In the general case use the above particular result on subintervals. We omit the details. ■

The next comparison lemma is owed to Lakshmikantham–Leela–Sivasundaram [34].

Lemma 12.2 *Let* $u, v \in C^1[a,b]$ *and* $F \in C([a,b] \times \mathbf{R})$. *Assume that*

$$u' \le F(t,u), \quad F(t,v) \le v',$$

$$F(t,z_1) - F(t,z_2) \le L(z_1 - z_2) \quad \textit{for } z_1 \ge z_2,$$

where $L \in \mathbf{R}_+$. *Then* $u(a) \le v(a)$ *implies* $u(t) \le v(t)$ *on* $[a,b]$.

Proof. Assume that $u(a) \le v(a)$. For small $\varepsilon > 0$ let

$$\widetilde{u} = u - \varepsilon\, e^{2L(t-a)}, \quad \widetilde{v} = v + \varepsilon\, e^{2L(t-a)}.$$

We have

$$\widetilde{u} < u, \quad v < \widetilde{v}, \quad \widetilde{u}(a) < \widetilde{v}(a).$$

In addition

$$\begin{aligned} \widetilde{u}' &= u' - 2\varepsilon\, Le^{2L(t-a)} \\ &\le F(t,u) - 2\varepsilon\, Le^{2L(t-a)} \\ &\le F(t,\widetilde{u}) + L\varepsilon\, e^{2L(t-a)} - 2\varepsilon\, Le^{2L(t-a)} \\ &< F(t,\widetilde{u}). \end{aligned}$$

Similarly

$$\widetilde{v}' > F(t,\widetilde{v}).$$

We claim that $\widetilde{u}(t) \le \widetilde{v}(t)$ on $[a,b]$, which proves the result as $\varepsilon \to 0$. Indeed, if not there exists a $t_0 \in (a,b]$ with

$$\widetilde{u}(t_0) = \widetilde{v}(t_0), \quad \widetilde{u}'(t_0) \ge \widetilde{v}'(t_0).$$

Then we obtain

$$\begin{aligned} F(t_0, \widetilde{u}(t_0)) &> \widetilde{u}'(t_0) \\ &\ge \widetilde{v}'(t_0) \\ &> F(t_0, \widetilde{v}(t_0)), \end{aligned}$$

a contradiction. ∎

Assume that $\varphi \in C[-\tau, 0]$ and let $u_0, v_0 \in C^1[0, t_1]$ be such that $u_0(0) = v_0(0) = \varphi(0)$ and $u_0(t) < v_0(t)$ on $(0, t_1]$. Let

$$\Omega = \{(t,z) : 0 < t \le t_1, \ u_0(t) < z < v_0(t)\}$$

and

$$\widetilde{\Omega} = \overline{\Omega} \cup \{(t, \varphi(t)) : -\tau \le t < 0\}.$$

Theorem 12.2 *Assume:*

(i) for $0 < t \le t_1$, *one has*

$$u_0'(t) \le f(t, u_0(t)) + g(t-\tau, \widetilde{u}_0(t-\tau)),$$

$$v_0'(t) \ge f(t, v_0(t)) + g(t-\tau, \widetilde{v}_0(t-\tau));$$

(ii) $f \in C(\overline{\Omega})$, $g \in C(\widetilde{\Omega})$, *the derivatives* f_z, f_{zz}, g_z *and* g_{zz} *exist and are continuous on* $\overline{\Omega}$, *and satisfy*

$$f_{zz} \ge 0, \quad g_z \ge 0, \quad g_{zz} \le 0 \quad on\ \overline{\Omega}. \tag{12.12}$$

Then there exist the sequences (u_k) *increasing and* (v_k) *decreasing which converge uniformly on* $[0, t_1]$ *to the unique solution* $u \in C^1[0, t_1]$ *of* (12.11) *satisfying* $u_0 \le u \le v_0$, *and the convergence is quadratic, i.e., there are constants* $a, b \in \mathbf{R}_+$ *such that*

$$|u_k - u|_\infty, \ |v_k - u|_\infty \le a\,|u_{k-1} - u|_\infty^2 + b\,|v_{k-1} - u|_\infty^2 \tag{12.13}$$

for all $k \ge 1$.

Proof. Notice (12.11) has a unique solution u satisfying $u_0 \le u \le v_0$; this follows by using the step method since $f(t,z)$ is Lipschitz in z on $\overline{\Omega}$.

In what follows we shall use the convexity of f and concavity of g by means of the inequalities

$$\begin{aligned} f(t,z) &\ge f(t,z') + f_z(t,z')(z - z'), \\ g(t,z) &\ge g(t,z') + g_z(t,z)(z - z'), \end{aligned} \tag{12.14}$$

which are true for all $(t,z), (t,z') \in \overline{\Omega}$. Suppose we have already built the functions u_k and v_k such that

$$u_0 \le u_k \le v_k \le v_0$$

and

$$\begin{aligned} u_k'(t) &\le f(t, u_k(t)) + g(t-\tau, \widetilde{u}_k(t-\tau)), \\ v_k'(t) &\ge f(t, v_k(t)) + g(t-\tau, \widetilde{v}_k(t-\tau)). \end{aligned} \tag{12.15}$$

Then we take $u_{k+1} = \alpha$ and $v_{k+1} = \beta$, α and β being the unique solutions of the following linear initial value problems with delay

$$\begin{cases} \alpha'(t) = F_k(t, \alpha(t), \widetilde{\alpha}(t-\tau)), & 0 < t \le t_1, \\ \alpha(0) = \varphi(0), \end{cases} \tag{12.16}$$

respectively

$$\begin{cases} \beta'(t) = G_k\left(t, \beta(t), \widetilde{\beta}(t-\tau)\right), & 0 < t \le t_1, \\ \beta(0) = \varphi(0), \end{cases} \tag{12.17}$$

where

$$\begin{aligned} F_k(t,x,y) &= f(t,u_k(t)) + f_z(t,u_k(t))(x - u_k(t)) \\ &\quad + g(t-\tau, \widetilde{u}_k(t-\tau)) \\ &\quad + g_z(t-\tau, \widetilde{v}_k(t-\tau))(y - \widetilde{u}_k(t-\tau)) \end{aligned} \tag{12.18}$$

and

$$\begin{aligned} G_k(t,x,y) &= f(t,v_k(t)) + f_z(t,u_k(t))(x - v_k(t)) \\ &\quad + g(t-\tau, \widetilde{v}_k(t-\tau)) \\ &\quad + g_z(t-\tau, \widetilde{v}_k(t-\tau))(y - \widetilde{v}_k(t-\tau)), \end{aligned} \tag{12.19}$$

where for $t \in [0,\tau]$, by $g_z(t-\tau, \varphi(t-\tau))$ we mean $g_z(0, \varphi(0))$. Notice for

$$u_k(t) \le x \le v_k(t), \quad \widetilde{u}_k(t-\tau) \le y \le \widetilde{v}_k(t-\tau),$$

(12.18), (12.14) and g_z decreasing yield

$$F_k(t,x,y) \le f(t,x) + g(t-\tau, y), \tag{12.20}$$

whilst (12.19), (12.14) and f_z increasing imply

$$G_k(t,x,y) \ge f(t,x) + g(t-\tau, y). \tag{12.21}$$

From (12.12), (12.14) and (12.15) we easily see that the following inequalities hold:

$$u_k'(t) \le F_k(t, u_k(t), \widetilde{u}_k(t-\tau)),$$

$$v_k'(t) \ge F_k(t, v_k(t), \widetilde{v}_k(t-\tau)).$$

We now successively apply Lemma 12.1 to the intervals $[0,\tau]$, $[\tau, 2\tau]$, ..., $[m\tau, t_1]$, where $m\tau < t_1 \le (m+1)\tau$. Thus we prove the existence of the solution $\alpha = u_{k+1}$ of (12.16) satisfying $u_k \le \alpha \le v_k$. Similarly we find a solution β of (12.17) such that $u_k \le \beta \le v_k$.

Further, by (12.20)–(12.21) one has

$$\begin{aligned} \alpha'(t) &\le f(t, \alpha(t)) + g(t-\tau, \widetilde{\alpha}(t-\tau)), \\ \beta'(t) &\ge f(t, \beta(t)) + g(t-\tau, \widetilde{\beta}(t-\tau)). \end{aligned} \tag{12.22}$$

Hence (12.15) also holds with $k+1$ instead of k. In order now to derive the inequality $\alpha \leq \beta$, that is $u_{k+1} \leq v_{k+1}$, we apply Lemma 12.2, successively on $[0,\tau]$, $[\tau, 2\tau]$, ..., $[m\tau, t_1]$, by taking

$$F(t,z) = F_0(t,z) + g(t-\tau, \widetilde{\alpha}(t-\tau)),$$

where

$$F_0(t,z) = \begin{cases} f(t,z) & \text{for } z \in [u_0(t), v_0(t)], \\ f(t,u_0(t)) & \text{for } z \leq u_0(t), \\ f(t,v_0(t)) & \text{for } z \geq v_0(t). \end{cases}$$

Suppose we have already proved that $\alpha(t) \leq \beta(t)$ for all $t \leq j\tau$. Then, for $t \in [j\tau, (j+1)\tau]$, we know that

$$\widetilde{\alpha}(t-\tau) \leq \widetilde{\beta}(t-\tau)$$

and since g is increasing (recall $g_z \geq 0$),

$$g(t-\tau, \widetilde{\beta}(t-\tau)) \geq g(t-\tau, \widetilde{\alpha}(t-\tau)).$$

Then by (12.22)

$$\begin{aligned} \alpha'(t) &\leq f(t,\alpha(t)) + g(t-\tau, \widetilde{\alpha}(t-\tau)), \\ \beta'(t) &\geq f(t,\beta(t)) + g(t-\tau, \widetilde{\alpha}(t-\tau)). \end{aligned}$$

Now Lemma 2 guarantees $\alpha(t) \leq \beta(t)$ on $[j\tau, (j+1)\tau]$ and so $\alpha(t) \leq \beta(t)$ for all $t \leq (j+1)\tau$. This argument applied successively yields $\alpha(t) \leq \beta(t)$ on $[0,t_1]$. Hence the sequences (u_k), (v_k) are well defined. Using standard arguments it is easy to conclude that they both converge to the unique solution u of problem (12.11).

Finally, we show that the convergence of the sequences (u_k) and (v_k) to the unique solution u of (12.11) is quadratic. Let

$$p_k(t) = u(t) - u_k(t), \quad q_k(t) = v_k(t) - u(t).$$

Note that $p_k(0) = q_k(0) = 0$ and $p_k(t) \geq 0$, $q_k(t) \geq 0$ for $t \in [0,t_1]$. Denote

$$\begin{aligned} |p_k|_{\infty,j} &= \max\{p_k(t) : t \in [0,(j+1)\tau] \cap [0,t_1]\}, \\ |q_k|_{\infty,j} &= \max\{q_k(t) : t \in [0,(j+1)\tau] \cap [0,t_1]\}, \end{aligned}$$

for $j = 0, 1, ..., m$. We shall prove that for each $j \in \{0,1,...,m\}$, there are $a_j, b_j \in \mathbf{R}_+$ such that

$$|p_{k+1}|_{\infty,j}, \ |q_{k+1}|_{\infty,j} \leq a_j |p_k|^2_{\infty,j} + b_j |q_k|^2_{\infty,j}. \tag{12.23}$$

Clearly, for $j = m$ (12.23) gives (12.13). Let $t \in [0, \tau]$. Then using the definition of u_{k+1} and the mean value theorem, we obtain

$$\begin{aligned} p'_{k+1}(t) &= f(t, u(t)) + g(t - \tau, \varphi(t - \tau)) - [f(t, u_k(t)) \\ &\quad + f_z(t, u_k(t))(u_{k+1}(t) - u_k(t)) + g(t - \tau, \varphi(t - \tau))] \\ &= f(t, u(t)) - [f(t, u_k(t)) + f_z(t, u_k(t))(u_{k+1}(t) - u_k(t))] \\ &= f_z(t, z_1) p_k(t) - f_z(t, u_k(t))(p_k(t) - p_{k+1}(t)) \\ &\leq f_{zz}(t, z_2) p_k^2(t) + f_z(t, u_k(t)) p_{k+1}(t) \\ &\leq M_2 |p_k|_{\infty,0}^2 + M_1 p_{k+1}(t), \end{aligned}$$

where $u_k(t) \leq z_2 \leq z_1 \leq u(t)$, $|f_z| \leq M_1$ and $|f_{zz}| \leq M_2$. It follows that for $t \in [0, \tau]$,

$$p_{k+1}(t) \leq M_2 \tau |p_k|_{\infty,0}^2 + M_1 \int_0^t p_{k+1}(s)\, ds.$$

Now the Gronwall's inequality implies

$$p_{k+1}(t) \leq M_2 \tau e^{M_1 \tau} |p_k|_{\infty,0}^2, \quad t \in [0, \tau].$$

Hence

$$|p_{k+1}|_{\infty,0} \leq M_2 \tau e^{M_1 \tau} |p_k|_{\infty,0}^2.$$

Similarly,

$$\begin{aligned} q'_{k+1}(t) &= f(t, v_k(t)) + f_z(t, u_k(t))(v_{k+1}(t) - v_k(t)) \\ &\quad + g(t - \tau, \varphi(t - \tau)) - [f(t, u(t)) + g(t - \tau, \varphi(t - \tau))] \\ &= f(t, v_k(t)) - f(t, u(t)) + f_z(t, u_k(t))(v_{k+1}(t) - v_k(t)) \\ &= f_z(t, z_1) q_k(t) + f_z(t, u_k(t))(q_{k+1}(t) - q_k(t)) \\ &\leq f_{zz}(t, z_2)(v_k(t) - u_k(t)) q_k(t) + f_z(t, u_k(t)) q_{k+1}(t) \\ &\leq d(q_k(t) + p_k(t)) q_k(t) + c q_{k+1}(t), \end{aligned}$$

where $u(t) \leq z_1 \leq v_k(t)$ and $u_k(t) \leq z_2 \leq z_1$. Hence

$$\begin{aligned} q'_{k+1}(t) &\leq M_1 q_{k+1}(t) + \frac{3}{2} M_2 q_k^2(t) + \frac{1}{2} M_2 p_k^2(t) \\ &\leq M_1 q_{k+1}(t) + \frac{3}{2} M_2 |q_k|_{\infty,0}^2 + \frac{1}{2} M_2 |p_k|_{\infty,0}^2. \end{aligned}$$

Using Gronwall's inequality again, we obtain

$$q_{k+1}(t) \leq \frac{3}{2} M_2 \tau e^{M_1 \tau} |q_k|_{\infty,0}^2 + \frac{1}{2} M_2 \tau e^{M_1 \tau} |p_k|_{\infty,0}^2, \quad t \in [0, \tau].$$

Thus
$$|q_{k+1}|_{\infty,0} \leq \frac{3}{2} M_2 \tau e^{M_1\tau} |q_k|^2_{\infty,0} + \frac{1}{2} M_2 \tau e^{M_1\tau} |p_k|^2_{\infty,0}.$$
Therefore (12.23) holds for $j = 0$, with $a_0 = M_2\tau e^{M_1\tau}$, $b_0 = \frac{3}{2} M_2 \tau e^{M_1\tau}$. Next, assume (12.23) holds for j. We shall prove (12.23) for $j+1$. Let $t \in [\tau, (j+1)\tau]$. We have

$$\begin{aligned}
p'_{k+1}(t) &= f(t,u(t)) + g(t-\tau, u(t-\tau)) - [f(t,u_k(t)) \\
&\quad + f_z(t,u_k(t))(u_{k+1}(t) - u_k(t)) + g(t-\tau, u_k(t-\tau)) \\
&\quad + g_z(t-\tau, v_k(t-\tau))(u_{k+1}(t-\tau) - u_k(t-\tau))] \\
&= f_z(t,z_1) p_k(t) + g_z(t-\tau, z_2) p_k(t-\tau) \\
&\quad + f_z(t,u_k(t))(p_{k+1}(t) - p_k(t)) \\
&\quad + g_z(t-\tau, v_k(t-\tau))(p_{k+1}(t-\tau) - p_k(t-\tau)) \\
&\leq [f_z(t,u(t)) - f_z(t,u_k(t))] p_k(t) \\
&\quad - [g_z(t-\tau, v_k(t-\tau)) - g_z(t-\tau, u_k(t-\tau))] p_k(t-\tau) \\
&\quad + f_z(t,u_k(t)) p_{k+1}(t) + g_z(t-\tau, v_k(t-\tau)) p_{k+1}(t-\tau) \\
&= f_{zz}(t,z_3) p_k^2(t) - g_{zz}(t-\tau, z_4)[v_k(t-\tau) - u_k(t-\tau)] \\
&\quad \times p_k(t-\tau) + f_z(t,u_k(t)) p_{k+1}(t) \\
&\quad + g_z(t-\tau, v_k(t-\tau)) p_{k+1}(t-\tau),
\end{aligned}$$

where $u_k(t) < z_1,\ z_3 < u(t),\ u_k(t-\tau) \leq z_2 \leq u(t-\tau)$. Also,

$$\begin{aligned}
&-g_{zz}(t-\tau, z_4)[v_k(t-\tau) - u_k(t-\tau)] p_k(t-\tau) \\
\leq\ & N_2 [q_k(t-\tau) + p_k(t-\tau)] p_k(t-\tau) \\
\leq\ & N_2 \left[\frac{3}{2} p_k^2(t-\tau) + \frac{1}{2} q_k^2(t-\tau)\right] \\
\leq\ & \frac{1}{2} N_2 \left[3 |p_k|^2_{\infty,j} + |q_k|^2_{\infty,j}\right].
\end{aligned}$$

Thus,

$$\begin{aligned}
p'_{k+1}(t) &\leq M_1 p_{k+1}(t) + M_2 |p_k|^2_{\infty,j+1} \\
&\quad + N_1 |p_{k+1}|_{\infty,j} + \frac{1}{2} N_2 \left[3 |p_k|^2_{\infty,j} + |q_k|^2_{\infty,j}\right],
\end{aligned}$$

where $|g_z| \leq N_1$, $|g_{zz}| \leq N_2$. Since

$$\begin{aligned}
|p_{k+1}|_{\infty,j} &\leq a_j |p_k|^2_{\infty,j} + b_j |q_k|^2_{\infty,j}, \\
|p_k|_{\infty,j} &\leq |p_k|_{\infty,j+1}, \\
|q_k|_{\infty,j} &\leq |q_k|_{\infty,j+1},
\end{aligned}$$

we obtain

$$p'_{k+1}(t) \le M_1 p_{k+1}(t) + \alpha |p_k|^2_{\infty,j+1} + \beta |q_k|^2_{\infty,j+1}.$$

Hence for $t \in [\tau, t_1]$ we have

$$\begin{aligned} p_{k+1}(t) &\le p_{k+1}(\tau) \\ &\quad + \left[\alpha |p_k|^2_{\infty,j+1} + \beta |q_k|^2_{\infty,j+1}\right](t_1 - \tau) + M_1 \int_\tau^t p_{k+1}(s)\, ds \\ &\le a_0 |p_k|^2_{\infty,0} + b_0 |q_k|^2_{\infty,0} + \left[\alpha |p_k|^2_{\infty,j+1} + \beta |q_k|^2_{\infty,j+1}\right](t_1 - \tau) \\ &\quad + M_1 \int_\tau^t p_{k+1}(s)\, ds \\ &\le \gamma |p_k|^2_{\infty,j+1} + \delta |q_k|^2_{\infty,j+1} + M_1 \int_\tau^t p_{k+1}(s)\, ds. \end{aligned}$$

Gronwall's inequality now yields the desired result

$$|p_{k+1}|_{\infty,j+1} \le a_{j+1} |p_k|^2_{\infty,j+1} + b_{j+1} |q_k|^2_{\infty,j+1},$$

for some constants a_{j+1}, $b_{j+1} \in \mathbf{R}_+$. Similarly,

$$\begin{aligned} q'_{k+1}(t) &= f(t, v_k(t)) \\ &\quad + f_z(t, u_k(t))(v_{k+1}(t) - v_k(t)) + g(t-\tau, v_k(t-\tau)) \\ &\quad + g_z(t-\tau, v_k(t-\tau))(v_{k+1}(t-\tau) - v_k(t-\tau)) \\ &\quad - f(t, u(t)) - g(t-\tau, u(t-\tau)) \\ &= f_z(t, z_1) q_k(t) + g_z(t-\tau, z_2) q_k(t-\tau) \\ &\quad + f_z(t, u_k(t))(q_{k+1}(t) - q_k(t)) \\ &\quad + g_z(t-\tau, v_k(t-\tau))(q_{k+1}(t-\tau) - q_k(t-\tau)) \\ &\le f_{zz}(t, z_3)(v_k(t) - u_k(t)) q_k(t) - g_{zz}(t-\tau, z_4) q_k^2(t-\tau) \\ &\quad + f_z(t, u_k(t)) q_{k+1}(t) + g_z(t-\tau, v_k(t-\tau)) q_{k+1}(t-\tau) \\ &\le f_{zz}(t, z_3)(q_k(t) - p_k(t)) q_k(t) - g_{zz}(t-\tau, z_4) q_k^2(t-\tau) \\ &\quad + f_z(t, u_k(t)) q_{k+1}(t) + g_z(t-\tau, v_k(t-\tau)) q_{k+1}(t-\tau), \end{aligned}$$

where $u(t) \le z_1 \le v_k(t)$, $u(t-\tau) \le z_2 \le v_k(t-\tau)$ and $u_k(t) \le z_3 \le z_1$. This yields

$$q'_{k+1}(t) \le M_1 q_{k+1}(t) + \alpha' |p_k|^2_{\infty,j+1} + \beta' |q_k|^2_{\infty,j+1}.$$

As above,

$$|q_{k+1}|_{\infty,j+1} \le a_{j+1} |p_k|^2_{\infty,j+1} + b_{j+1} |q_k|^2_{\infty,j+1},$$

eventually, with some new constants $a_{j+1}, b_{j+1} \in \mathbf{R}_+$. Thus (12.23) also holds for $j+1$. Therefore (12.23) is true for all $j \in \{0, 1, ..., m\}$. ■

Corollary 12.1 *Assume* $\varphi \in C[-\tau, 0]$, $u_0, v_0 \in C^1[0, t_1]$, $u_0(0) = v_0(0) = \varphi(0)$, $u_0 < v_0$ *on* $(0, t_1]$, $f \in C(\widetilde{\Omega})$, *and*

$$\begin{aligned} u_0'(t) &\leq f(t, u_0(t)) - f(t-\tau, \widetilde{u}_0(t-\tau)), \\ v_0'(t) &\geq f(t, v_0(t)) - f(t-\tau, \widetilde{v}_0(t-\tau)). \end{aligned}$$

If the derivatives f_z *and* f_{zz} *exist, are continuous on* $\overline{\Omega}$, *and*

$$f_z \leq 0, \quad f_{zz} \geq 0 \quad \text{on } \overline{\Omega},$$

then there exist the sequences (u_k) *increasing and* (v_k) *decreasing which converge uniformly and quadratically on* $[0, t_1]$ *to the unique solution* u *of* (12.10) *satisfying* $u_0 \leq u \leq v_0$.

An interesting problem is to obtain nonsmooth variants of the quasilinearization methods in terms of divided differences instead of derivatives. A partial result on this point can be found in Balázs–Muntean [5] and Goldner–Trîmbiţaş [24]. An extension of the generalized quasilinearization method in order to obtain convergence of order $p > 2$ is due to Deo–McGloin Knoll [21]. Another interesting problem is to express the conditions of such type of results, in terms of convex or quasiconvex functions of superior order (see Popoviciu [43], Popoviciu [42] and Cristescu [16]). A general theory in ordered Banach spaces of both monotone iterative technique and generalized quasilinearization method with applications to semilinear problems is developed in the forthcoming paper by Buică–Precup [9].

References: Part III

[1] AGARWAL, R.P. and O'REGAN, D., Fixed point theory for generalized contractions on spaces with two metrics, *J. Math. Anal. Appl.* **248** (2000), 402–414.

[2] AMMAN, H., Fixed point equations and nonlinear eigenvalues problems in ordered Banach spaces, *SIAM Review* **18** (1976), 620–709.

[3] AVRAMESCU, C., Sur l'existence des solutions convergentes pour des équations intégrales, *An. Univ. Craiova Ser. a V-a,* no. 2 (1974), 87–98.

[4] BAINOV, D. and HRISTOVA, S., Monotone-iterative techniques of V. Lakshmikantham for a boundary value problem for systems of integro-differential equations, *Math. J. Toyama Univ.* **18** (1995), 169–178.

[5] BALÁZS, M. and MUNTEAN, I., A unification of Newton's methods for solving equations, *Mathematica (Cluj)* **21** (44) (1979), 117–122.

[6] BELLMAN, R. and KALABA, R., *Quasilinearization and Nonlinear Boundary Value Problems,* American Elsevier, New York, 1965.

[7] BREZIS, H., *Analyse Fonctionnelle,* Masson, Paris, 1983.

[8] BUICĂ, A., Existence results for evolution equations via monotone iterative techniques, *Dynam. Contin. Discrete Impuls. Systems,* to appear.

[9] BUICĂ, A. and PRECUP, R., Abstract generalized quasilinearization method for coincidences, to appear.

[10] CABADA, A. and NIETO, J.J., Quasilinearization and rate of convergence for higher-order nonlinear periodic boundary-value problems, *J. Optim. Theory Appl.* **108** (2001), 97–107.

[11] CAÑADA, A., Method of upper and lower solutions for nonlinear integral equations and an application to an infectious disease model. In: *Dynamics of Infinite Dimentional Systems* (S.N. Chow and J.K. Hale eds.), Springer, Berlin, 1987, 39–44.

[12] CAÑADA, A. and ZERTITI, A., Method of upper and lower solutions for nonlinear delay integral equations modelling epidemics and population growth, *Math. Models and Methods in Applied Sciences* **4** (1994), 107–120.

[13] CARL, S. and HEIKKILÄ, S., Operator and differential equations in ordered spaces, *J. Math. Anal. Appl.* **234** (1999), 31–54.

[14] CARL, S. and LAKSHMIKANTHAM, V., Generalized quasilinearization for quasilinear parabolic equations with nonlinearities of DC type, *J. Optimization Theory Appl.* **109** (2001), 27–50.

[15] CONSTANTIN, A., Monotone iterative technique for a nonlinear integral equation, *J. Math. Anal Appl.* **205** (1997), 280–283.

[16] CRISTESCU, G., E-convexity of superior order of functionals on a metric space, *Bull. Appl. Comput. Math. (Budapest)* **92** (2000), 65–72.

[17] CRISTESCU, R., *Ordered Vector Spaces and Linear Operators*, Ed. Academiei – Abacus Press, Bucureşti – Kent, 1976.

[18] DADS, A., EZZINBI, K. and ARINO, O., Positive almost periodic solution for some nonlinear delay integral equation, *Nonlinear Studies* **3** (1996), 85–101.

[19] De COSTER, C. and HABETS, P., Existence and multiplicity of positive solutions of the Ginzburg–Landau boundary value problem, *J. Comput. Appl. Math.* **113** (2000), 317–327.

[20] DEIMLING, K., *Nonlinear Functional Analysis*, Springer–Velag, Berlin–Heidelberg–New York–Tokyo, 1985.

[21] DEO, S.G. and McGLOIN KNOLL, C., kth order convergence of an iterative method for integro-differential equations, *Nonlinear Studies* **5** (1998), 191–200.

[22] FABRY, CH. and HABETS, P., Upper and lower solutions for second-order boundary value problems with nonlinear boundary conditions, *Nonlinear Anal.* **10** (1986), 985–1007.

[23] FRIGON, M., Fixed point results for generalized contractions in gauge spaces and applications, *Proc. Amer. Math. Soc.* **128** (2000), 2957–2965.

[24] GOLDNER, G. and TRÎMBIŢAŞ, R., A combined method for a two-point boundary value problem, *Pure Math. Appl.* **11** (2000), 255–264.

[25] GRANAS, A., Continuation method for contractive maps, *Topol. Methods Nonlinear Anal.* **3** (1994), 375–379.

[26] GRANAS, A., GUENTHER, R. and LEE, J., Nonlinear boundary value problems for ordinary differential equations, *Dissertationes Math.* **244,** PWN, Warsaw, 1985.

[27] GUO, D. and LAKSHMIKANTHAM, V., *Nonlinear Problems in Abstract Cones,* Academic Press, Boston, 1988.

[28] GUO, D. and LAKSHMIKANTHAM, V., Positive solutions of nonlinear integral equations arising in infectious diseases, *J. Math. Anal. Appl.* **134** (1988), 1–8.

[29] GUO, D., LAKSHMIKANTHAM, V. and LIU, X., *Nonlinear Integral Equations in Abstract Spaces,* Kluwer Academic Publishers, Dordrecht–Boston–London, 1996.

[30] HEIKKILÄ, S. and LAKSHMIKANTHAM, V., *Monotone Iterative Techniques for Discontinuous Nonlinear Differential Equations,* Marcel Dekker, New York, 1994.

[31] JANKÓ, B., The *Solving of the Nonlinear Operator Equations in Banach Spaces* (Romanian), Ed. Academiei, Bucureşti, 1969.

[32] KANTOROVITCH, L. and AKILOV, G., *Analyse Fonctionnelle,* Mir Publishers, Moscow, 1981.

[33] KRASNOSELSKII, M.A., *Positive Solutions of Operator Equations,* Noordhoff, Groningen, 1964.

[34] LAKSHMIKANTHAM, V., LEELA, S. and SIVASUNDARAM, S, Extensions of the method of quasilinearization, *J. Optim. Theory Appl.* **87** (1995), 379–401.

[35] LIZ, E. and NIETO, J.J., Periodic boundary value problem for integrodifferential equations with general kernel, *Dynam. Systems Appl.* **3** (1994), 297–304.

[36] MOORE, R.E., *Computational Functional Analysis,* Ellis Horwood Ltd., Chichester; Halsted Press, Wiley, New York, 1985.

[37] MUREŞAN, V. and TRIF, D., Newton's method for nonlinear differential equations with linear deviating argument, *Studia Univ. Babeş–Bolyai Math.* **41**, no. 4 (1996), 89–95.

[38] O'REGAN, D. and PRECUP, R., *Theorems of Leray–Schauder Type and Applications,* Gordon and Breach Science Publishers, Amsterdam, 2001.

[39] O'REGAN, D. and PRECUP, R., Continuation theory for contractions on spaces with two vector-valued metrics, to appear.

[40] PEROV, A.I. and KIBENKO, A.V., On a certain general method for investigation of boundary value problems (Russian), *Izv. Akad. Nauk SSSR* **30** (1966), 249–264.

[41] PICCININI, L.C., STAMPACCHIA, G. and VIDOSSICH, G., *Ordinary Differential Equations in* $\mathbf{R}^n$, Springer–Verlag, New York–Berlin–Heidelberg–Tokyo, 1984.

[42] POPOVICIU, E., *Sur une allure de quasi-convexité d'ordre supérieure,* Rev. Anal. Numér. Théor. Approx. **11**, (1982), 129–137.

[43] POPOVICIU, T., *Les Fonctions Convexes,* Hermann & Cie, Paris, 1944.

[44] POTRA, F.A. and RHEINBOLDT, W.C., On the monotone convergence of Newton's method, *Computing* **36** (1986), 81–90.

[45] POUSO, R.L., Nonordered discontinuous upper and lower solutions for first-order ordinary differential equations, *Nonlinear Anal.* **45** (2001), 391–406.

[46] PRECUP, R., Periodic solutions for an integral equation from biomathematics via Leray–Schauder principle, *Studia Univ. Babeş–Bolyai Math.* **39**, No. 1 (1994), 47–58.

[47] PRECUP, R., Monotone technique to the initial values problem for a delay integral equation from biomathematics, *Studia Univ. Babeş–Bolyai, Math.* **40**, No. 2 (1995), 63–73.

[48] PRECUP, R., Convexity and quadratic monotone approximation in delay differential equations. In: *Proc. Scientific Meeting of the "Aurel Vlaicu" Univ. of Arad*, Univ. "Aurel Vlaicu" Arad, 1997, 153–158.

[49] PRECUP, R., Analysis of some neutral delay differential equations, *Studia Univ. Babeş-Bolyai Math.* **44**, No.3 (1999), 67–84.

[50] PRECUP, R., Discrete continuation method for boundary value problems on bounded sets in Banach spaces, *J. Comput. Appl. Math.* **113** (2000), 267–281.

[51] PRECUP, R., Discrete continuation method for nonlinear integral equations in Banach spaces, *Pure Math. Appl.* **11** (2000), 375–384.

[52] PRECUP, R., Nonlinear evolution equations via the discrete continuation method. In: *Proc. of the "Tiberiu Popoviciu" Itinerant Seminar of Functional Equations, Approximation and Convexity* (E. Popoviciu ed.), Srima, Cluj–Napoca, 2000, 187–192.

[53] PRECUP, R., Continuation results for mappings of contractive type, *Seminar on Fixed Point Theory Cluj–Napoca* **2** (2001), 23–40.

[54] PRECUP, R., The continuation principle for generalized contractions, *Bull. Appl. Comput. Math. (Budapest)*, **96**-C (2001), 367–373.

[55] RUS, I.A., *Principles and Applications of the Fixed Point Theory* (Romanian), Dacia, Cluj–Napoca, 1979.

[56] RUS, I.A., *Generalized Contractions and Applications,* Cluj University Press, Cluj–Napoca, 2001.

[57] SCHMIDT, J.W. and SCHNEIDER, H., Monoton einschließende Verfahren bei additiv zerlegbaren Gleichungen, *Z. Angew. Math. Mech.* **63** (1983), 3–11.

[58] ŞERBAN, M.A., Existence and uniqueness theorems for positive solutions of Chandrasekhar's equation, *Mathematica (Cluj)* **41** (64) (1999), 91–103.

[59] TSACHEV, T. and ANGELOV, V.G., Fixed points of nonself-mappings and applications, *Nonlinear Anal.* **21** (1993), 9–16.

[60] VANDERGRAFT, J.S., Newton's method for convex operators in partially ordered spaces, *SIAM J. Numer. Anal.* **4** (1967), 406–432.

[61] WANG, M.-X., CABADA, A. and NIETO, J.J., Monotone method for nonlinear second order periodic boundary value problems with Carathéodory functions, *Ann. Polon. Math.* **58** (1993), 221–235.

Index

9 781402 008443